Contents

Specification matching grid

		AS TOPICS		A2 TOPICS	
Chapter	*Topic*	*AQA*	*Edexcel*	*AQA*	*Edexcel*
1	Cells	Module 1 (10.1) Module 1 (10.2) Module 1 (10.3)	Unit 1 (1.3)		Unit 4 Option A (A.1)
2	Biological molecules	Module 1 (10.4)	Unit 1 (1.1)		
3	Enzymes	Module 1 (10.5)	Unit 1 (1.2)		
4	Breathing and gas exchange	Module 1 (10.6)	Unit 2 (H.1)		
5	Blood and circulation	Module 1 (10.6) Module 1 (10.7) Module 1 (10.8)		Module 7 (16.6)	Unit 4 Option C (C.1)
6	The body and exercise	Module 1 (10.8)		Module 7 (16.10)	Unit 4 Option C (C.2)
7	Cell division		Unit 1 (1.4)		
8	DNA and protein synthesis		Unit 1 (1.4)		
9	Gene technology				Unit 5 (H.1)
10	Parasites, pathogens and disease	Module 3 (12.1) Module 3 (12.2)		Module 5 (14.9) Module 5 (14.10)	Unit 4 Option C (C.3)
11	Heart disease	Module 1 (10.8) Module 3 (12.7)			Unit 4 Option C (C.3)
12	Immunity, diagnosis and disease	Module 3 (12.3)			Unit 4 Option C (C.1)
13	Inheritance			Module 5 (14.1)	Unit 5 (H.1)
14	Evolution			Module 5 (14.2) Module 5 (14.3)	Unit 5 (H.1) Unit 5 (H.2)
15	Photosynthesis		Unit 3 (3.1) Unit 3 (3.3)	Module 5 (14.6) Module 5 (14.7)	
16	Respiration			Module 5 (14.8)	Unit 4.1
17	Ecology		Unit 3 (3.2) Unit 3 (3.4) Unit 3 (3.6)	Module 5 (14.4) Module 5 (14.5) Module 5 (14.9)	Unit 5 (H.4)
18	Humans and environment		Unit 3 (3.4)		
19	Reproduction		Unit 2H (2H.4)	Module 7 (16.1) Module 7 (16.2)	
20	Lifestages		Unit 2H (2H.4)	Module 7 (16.3) Module 7 (16.12)	Unit 5 (5H.3)
21	Digestion and nutrition		Unit 2H (2H.1) Unit 3 (3.1)		Unit 4 Option B (B.1)
22	Homeostasis		Unit 2H (2H.3)	Module 7 (16.11)	Unit 4 (4.2)
23	Nervous system			Module 7 (16.7) Module 7 (16.8) Module 7 (16.10)	Unit 4 (4.2)
24	Effectors: glands and muscles			Module 7 (16.9)	Unit 4 (4.2) Unit 4 Option C (C.1)

How to use this book . . .

The introduction to each chapter gives a brief outline of the topics and ideas which can be practised by working through the questions. A good basic level of science knowledge and ideas is assumed.

A wide variety of question styles offers you the opportunity to practise different skills, including:

- basic factual recall
- application of knowledge
- data interpretation and manipulation
- understanding experimental method
- numeracy
- writing accounts.

Using this book regularly as a course companion will reinforce your understanding of important concepts, as well as testing the learning from your taught sessions. It can also be used as a comprehensive revision tool. Answers are provided to help you learn independently, and to allow you to work at your own pace. Bear in mind that the mark schemes given only serve as an indication of how points might be allocated at examination.

We hope that using this book will help you through to a successful conclusion to your course. Good luck!

Jane Vellacott and Sarah Side

1 Cells

Cells make up all living organisms. The simplest are bacteria, which have prokaryotic cells. These are small, simple and lack a nucleus. Plants, animals and fungi are composed of larger eukaryotic cells which contain a nucleus and a variety of membrane-bound organelles such as mitochondria. All cells are surrounded by a cell membrane which is important in controlling the passage of molecules and ions in and out of the cell by active processes. Plant cells are bound by a cellulose cell wall. Metabolism, the synthesis and catabolism of larger molecules, is controlled in the cell by enzymes. During development, cells become differentiated and organised into tissues such as muscle, bone and blood in order to carry out specialised functions.

1.1 The prokaryotic cell

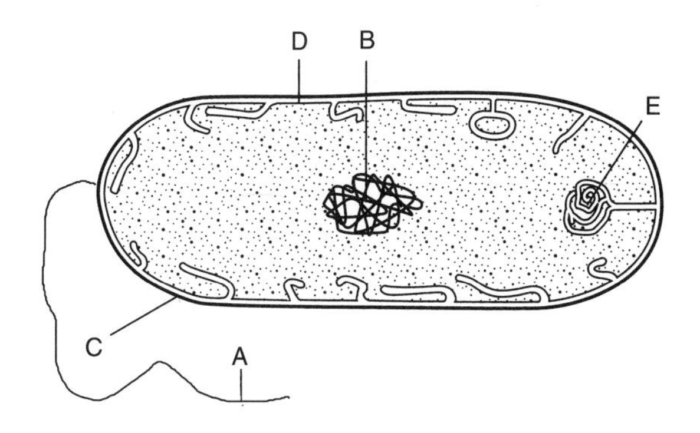

Figure 1.1 A bacterial cell

(a) Label A–E. **[5]**

(b) Unlike a plant cell wall the bacterial cell wall is <u>not</u> made of cellulose. What is it composed of? **[1]**

(c) Name two food materials that might be stored in a cell such as this. **[2]**

(d) Pathogenic bacteria are often covered by a thick capsule of mucopolysaccharide. Suggest a reason for this. **[1]**

(e) Bacterial cells divide rapidly. Within a few hours the group of millions of cells may be visible to the naked eye. What is it then called? **[1]**

1.2 (a) Name one example of a prokaryotic cell and one of a eukaryotic cell. **[2]**

(b) The table compares features of prokaryotic and eukaryotic cells. If the feature is present put a tick in the appropriate box, but if it is absent put a cross in the box.

Feature	Prokaryotic cell	Eukaryotic cell
Nucleus		
Mitochondria		
Mesosome		
Ribosome		
Cell surface membrane		
Endoplasmic reticulum		
Plasmid		

[7]

1.3 Cell structure

The electron micrograph shows an animal cell. It is magnified × 2000.

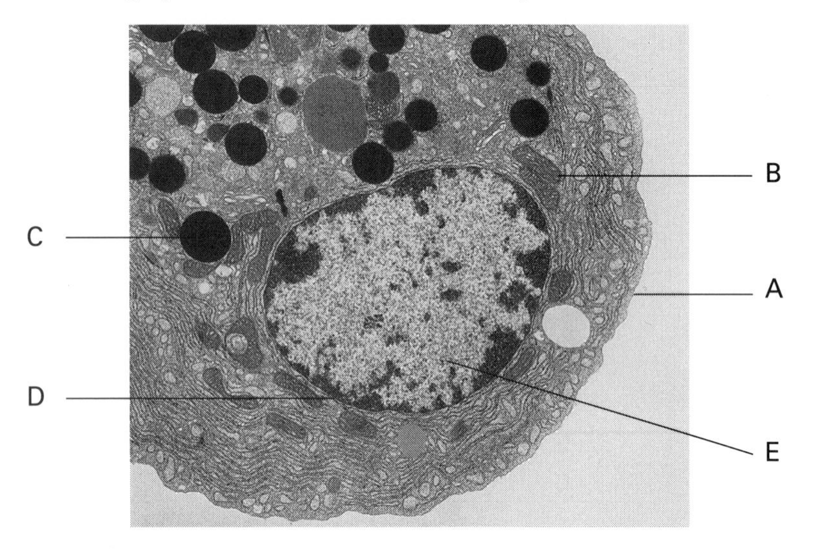

Figure 1.2 Part of an animal cell

(a) Label structures A–E. **[6]**

(b) Work out the actual length of structure B in μm. Show your working. **[2]**

(c) Explain the functions of the following: **(i)** cell membrane **(ii)** mitochondria **(iii)** Golgi body **(iv)** lysosomes **(v)** ribosomes **(vi)** endoplasmic reticulum. **[6]**

(d) Write a short paragraph to explain the structural differences between the cell shown above and a palisade cell from mesophyll tissue in a leaf. **[3]**

1.4 What is a tissue? Explain your answer in a short account with reference to suitable tissues found in humans. **[25]**

1.5 Microscopy

(a) Copy this passage about electron microscopy using suitable terms to fill the blanks.

Electron microscopes can ___________ by at least 500 000 times. They have much greater ___________ power than the light or optical microscope. This means that they reveal greater detail because they can distinguish two objects that are very ___________ together. The radiation source used is ___________ focused on to the specimen by ___________ . To prevent the radiation scattering there is a ___________ inside the electron microscope.

A ___________ electron microscope shows the internal structure of an object and a ______________________ shows a 3-dimensional image of the surface. The image of the specimen projected on to a fluorescent screen or on to a photographic plate is called an ________________. **[9]**

(b) Give three advantages of using a light microscope and three disadvantages of using an electron microscope. **[6]**

(c) Explain why great care must be taken in interpreting the image obtained by an electron microscope. **[1]**

1.6 Cell fractionation is a technique used to isolate cell components in a state in which they can be studied further. The process is summarised in the following flow diagram.

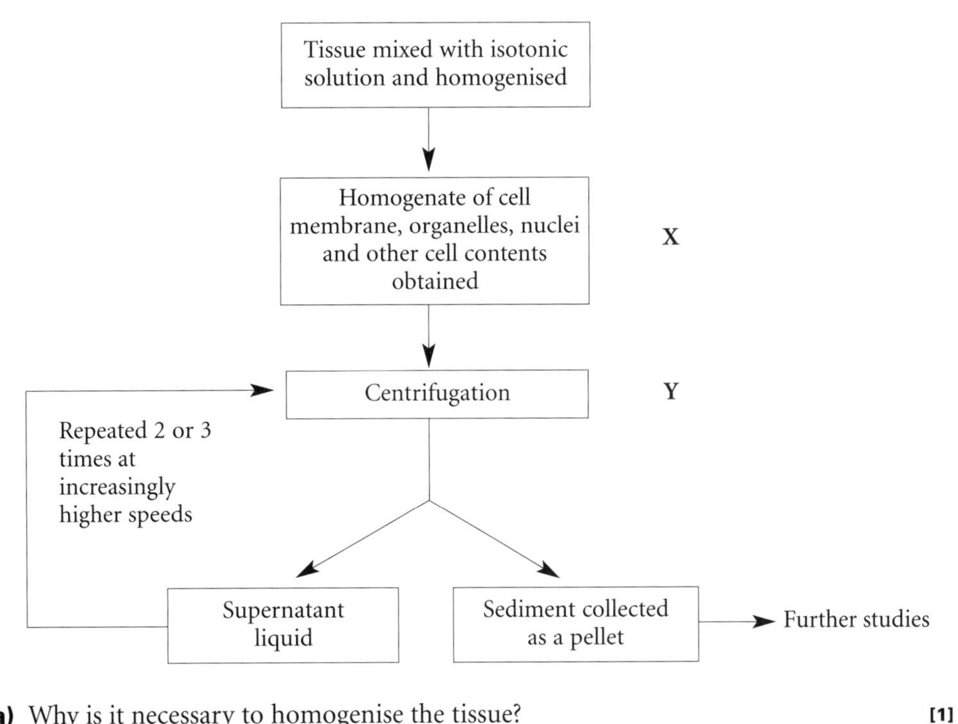

(a) Why is it necessary to homogenise the tissue? **[1]**

(b) Why is isotonic saline or sugar solution added to the tissue? **[1]**

(c) What is centrifugation? Why is it necessary to repeat centrifugation on successive samples of supernatant at increasing speeds? **[2]**

(d) Why are cell components separated by this method rather than by filtration? **[1]**

(e) Draw and label a test tube and its contents as it would appear when removed from the centrifuge at stage Y. **[2]**

(f) Suggest one other component found in the homogenate at stage X. **[1]**

(g) Cell fractionation is carried out at low temperature. Suggest a reason for this. **[1]**

(h) Ribosomes, chloroplasts, mitochondria and nuclei can all be obtained from cells by fractionation. List them in order of decreasing density. **[1]**

1.7 The cell membrane

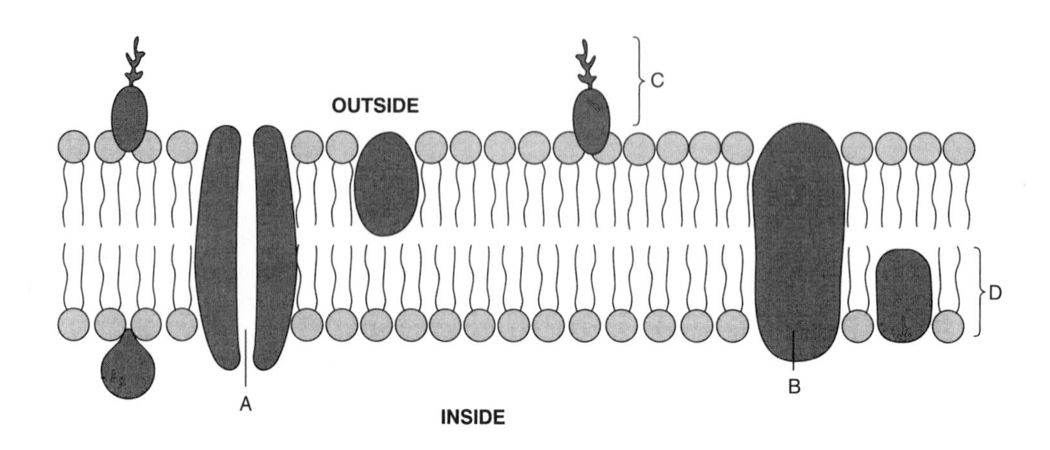

Figure 1.3 Structure of the cell surface membrane.

(a) Approximately how thick is the cell surface membrane? **[1]**

(b) Label structures A B C D. **[4]**

(c) Biologists use a model called the 'fluid mosaic model', to describe the structure of the cell surface membrane. Explain why the model has this name. **[2]**

(d) Explain how phospholipid molecules are arranged in the cell surface membrane. **[2]**

(e) Name one other lipid molecule found in the cell surface membrane. **[1]**

(f) Which molecules can pass through the phospholipid layer of the cell membrane? **[1]**

(g) Explain the terms extrinsic protein, intrinsic protein, glycoprotein. **[3]**

(h) Give three functions for proteins in the cell surface membrane. **[3]**

(i) Suggest why membranes with relatively high cholesterol content are less permeable to water than those with a lower cholesterol content. **[1]**

1.8 The movement of molecules

(a) Give three reasons why molecules or ions must move through cell membranes. **[3]**

(b) List five methods by which particles can move through the cell surface membrane, indicating for each, whether it is an active or a passive process. **[5]**

(c) Define diffusion. Name three examples of sites where it occurs in the human body. **[4]**

(d) What is a disadvantage of diffusion? **[1]**

(e) Complete the blanks in the following passage about how glucose molecules diffuse through the cell surface membrane.

Glucose molecules cannot dissolve in the _______________ of the cell membrane. They may pass through _______________ . A second method is to move by _______________ diffusion. Some carrier _______________ have a specific _______________ site to which glucose molecules can attach. Having picked up a molecule the carrier then changes_______________ and deposits the glucose on the other _______________ of the cell membrane. This process relies only on the _______________ energy of the glucose molecules and does not require _______________ from the cell, and so it is a _______________ process. **[10]**

1.9 Osmosis

Figure 1.4 shows an experiment that was set up in a school laboratory.

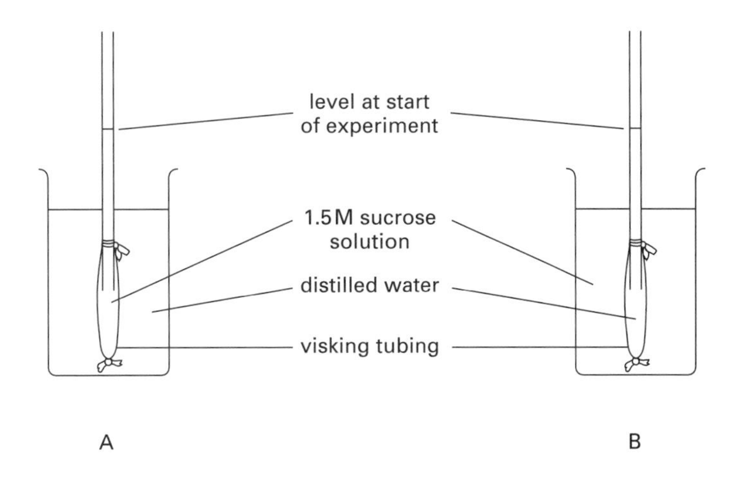

Figure 1.4

(a) Visking tubing is a differentially (partially) permeable membrane. What does this mean? **[1]**

(b) What changes would you expect to have taken place in A and B after 30 minutes? Explain your answer. **[6]**

(c) Why is osmosis 'a special case of diffusion'? **[1]**

(d) Water potential, solute potential and pressure potential are all terms used when describing the water relations of plant cells. Match each term to one of the following statements.

A The potential of a solution to take in water by osmosis, due to the solute dissolved in it.

B The tendency of water to move out of a solution by osmosis.

C The pressure exerted by the cell vacuole and cell contents on the cell wall. **[3]**

(e) Write a word equation and beneath it a corresponding symbol equation, to show the connection between water potential, solute potential and pressure potential. **[2]**

(f) Distinguish between the following:
hypotonic, hypertonic and isotonic solutions. **[3]**

(g) Delete the incorrect words or figures in the following sentences:

(i) The water potential of pure water at standard atmospheric pressure is 0/100 atmospheres/kPa.

(ii) The water potential of a solution always has a positive/negative value.

(iii) Water potential is higher/lower in a hypertonic solution than in a hypotonic solution.

(iv) A solution with water potential –320 kPa has a higher/lower water potential than a solution of –560 kPa.

(v) Solution F is separated from solution G by a differentially (partially) permeable membrane. In Solution F $\psi = -480$ kPa and in solution G $\psi = -170$ kPa. Water moves by osmosis from F to G/G to F. **[6]**

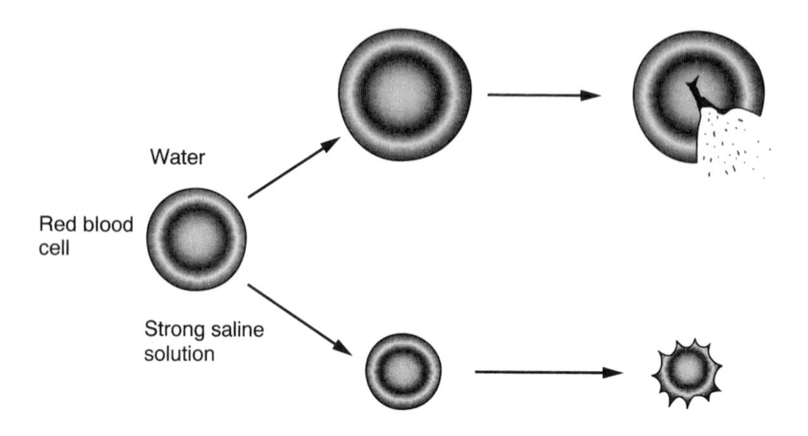

Figure 1.5

(h) Figure 1.5 shows what happens to red blood cells when they are put into either water or saline solution. Explain fully what has happened in the cells. **[4]**

(i) Make three labelled drawings to show what happens to a turgid plant cell that is placed in strong sucrose solution. **[3]**

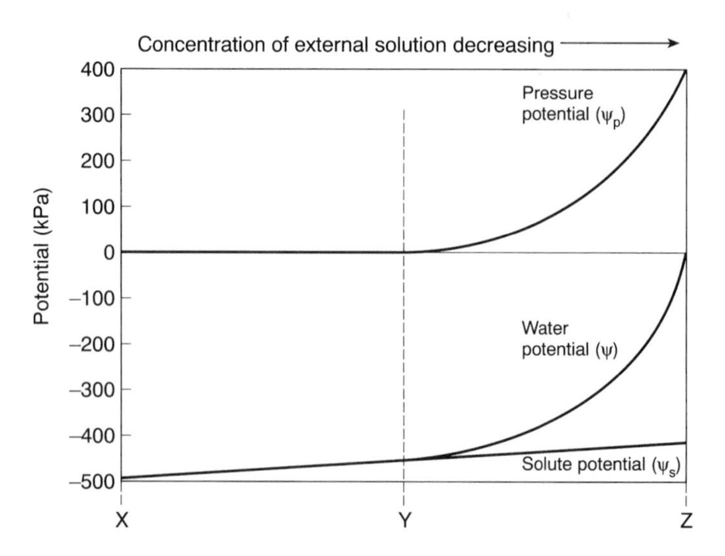

Figure 1.6

(j) Look at the graph in Figure 1.6.
X, Y and Z represent three cells. One cell is fully turgid, one is fully plasmolysed and the third shows incipient plasmolysis.
(i) Which letter represents which cell?
(ii) Which cells, if any, have a pressure potential of zero? **[4]**

1.10 Active transport

(a) Put the following statements in order to write a paragraph about active transport.

A Movement depends upon intrinsic proteins in the cell membrane specific to particular ions or molecules.
B The particle is then released at the other side of the membrane and the carrier protein reverts to its original shape.
C These form the 'biological' or protein pumps.
D A molecule of ATP attaches to a second site on the carrier.
E Active transport is a mechanism for moving molecules or ions against a concentration gradient.
F The ATP is hydrolysed causing the conformation of the carrier protein to change.
G The particle to be moved binds to the protein carrier. **[7]**

(b) Name one site in the human body of :
(i) a sodium-potassium pump
(ii) a proton pump
and say briefly what each does. **[4]**

(c) Draw a labelled diagram of a cell to explain the terms endocytosis and exocytosis. **[4]**

(d) How are endocytosis and exocytosis similar to active transport? **[2]**

1.11 The following table compares ways in which particles move. Copy the table and put a tick if the statement is correct and a cross if it is incorrect.

	Active transport	Osmosis	Facilitated diffusion	Diffusion
Occurs in living cells only				
Particles travel from high to low concentration				
Uses protein carriers				
ATP provides the source of energy				

The complete row must be correct to gain a mark. **[4]**

1.12 *Every winter people are exposed to the viral disease influenza. To the vulnerable, this is potentially a life threatening condition. The most devastating outbreak to date was the pandemic of Spanish flu in 1918 when 20 million people died. Some protection is now available from vaccines which contain chemically inactivated influenza virus. However these do not afford 100% protection against the disease.*

Influenza virus, a retrovirus, is unusually complex because it consists of eight separate pieces of RNA and numerous proteins. For the first time, scientists in Wisconsin have engineered a flu virus in the laboratory. They introduced all the pieces of the influenza A virus into human embryonic kidney cells using plasmids. The cells then replicated the RNA and proteins necessary in a flu virus. Up to 50 million virions were produced from each millilitre of cell culture.

The hope is that in the future a weakened virus will be designed from which a 'master vaccine' can be manufactured. This will be modified quickly to work against new strains of influenza virus as they arise.

(a) Suggest three groups of people to whom influenza may be potentially life threatening. **[3]**

(b) Briefly describe the structure of a virus particle. (You may include a labelled diagram). **[4]**

(c) What is a retrovirus? Name one example other than influenza. **[2]**

(d) Why do vaccines against influenza fail to offer 100% protection? **[1]**

(e) Explain the terms 'plasmid' and 'virion'. **[2]**

(f) How do new strains of influenza virus arise? **[1]**

(g) Why is there an incubation period of 48–72 hours between being infected with the influenza virus and developing symptoms of the disease? **[2]**

2 Biological molecules

Biological molecules are polymers, chains of smaller molecules joined during condensation reactions. Hydrolysis reactions break bonds between monomers by the addition of water. Proteins, carbohydrates, lipids and nucleic acids all contain carbon and so are known as organic molecules. They also contain hydrogen and oxygen. Each has its own physical and chemical characteristics and plays a vital part in the metabolism of all living organisms. Proteins are used in the growth, replacement and repair of cells; enzymes are proteins. Carbohydrates and lipids are important energy sources and play a part in cell structure. Water is essential to all forms of life because of its special physical properties.

2.1 Properties of water

(a) Which of the following is the correct structure of the water molecule? **[1]**

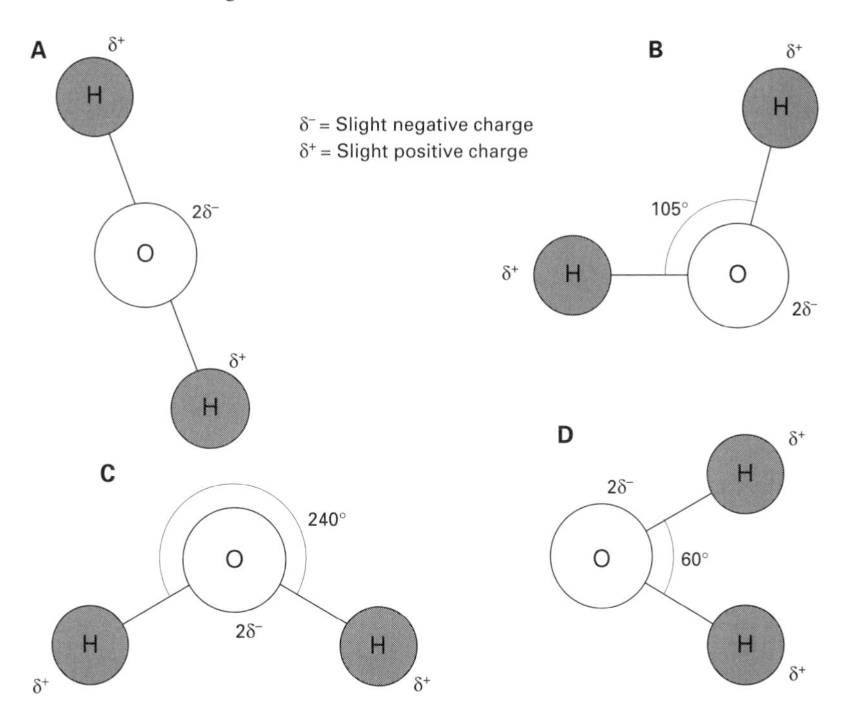

Figure 2.1

(b) Explain what is meant by the phrase, water molecules have 'polarity'. **[1]**

(c) Draw a labelled diagram to show the formation of a hydrogen bond between two water molecules. What is the significance of this? **[3]**

(d) Which of the following are soluble in water:
fatty acids, sodium chloride (NaCl/common salt), glucose, phospholipids, proteins, starch, amino acids? **[3]**

(e) Why is water a solvent for polar molecules? **[2]**

(f) Complete the following statements that refer to the properties of water by deleting incorrect words and filling blanks:

(i) Specific heat capacity is the energy required to raise the temperature of 1kg water by 1°C. Water in oceans, lakes and other water bodies maintains its temperature in spite of fluctuations in air temperature, due to its high/low specific heat.

(ii) Surface tension, the force of attraction between neighbouring surface molecules, causes the surface of a liquid to expand/contract. Water occupies the lowest possible surface area because it has high/low surface tension.

(iii) When liquids change to solid state they usually become less/more dense. This is true/false for water. Water is most dense at 0°C/4°C/100°C. This is important to aquatic organisms because ________________ .

(iv) Water is transparent/opaque in the liquid state. This allows aquatic plants to carry out respiration/photosynthesis.

(v) Liquids can change state to vapour by evaporation/condensation. The energy required to do this is called the specific latent heat of vaporisation. A person getting out of a swimming pool may feel cold because water has a high/low latent heat of vaporisation. What is happening is ________________ . **[12]**

2.2 Write an account of how water plays a vital role in the functioning of the human body. **[10]**

2.3 The following terms are essential to understanding the chemistry of biological systems.

(a) Define and give an example for each term :

A condensation
B hydrolysis
C monomer
D polymer
E anabolism
F catabolism. **[12]**

(b) Copy and complete the following table:

	Polysaccharides	**Lipids**	**Proteins**
Elements present in all molecules		carbon, hydrogen, oxygen	
Subunits	hexose sugars		
Bonds between subunits			peptide
Are they found in cell surface membranes?		yes	
Function			growth, repair
Are they stored in human tissues?		yes (as fat)	
Can they act as a human respiratory substrate?	yes (converted first to glucose)		

[14]

2.4 Carbohydrates

(a) List three chemical characteristics of monosaccharides. **[3]**

(b) Name the groups of monosaccharides that have 3, 5 and 6 atoms of carbon in each molecule. **[3]**
Give the name and chemical formula of the monosaccharide found in **(i)** RNA and **(ii)** DNA. How do these two sugars differ from one another? **[5]**

(c) What is meant by the term chemical 'isomer'? **[1]**
Give three examples of hexose sugars which are chemical isomers. **[1]**

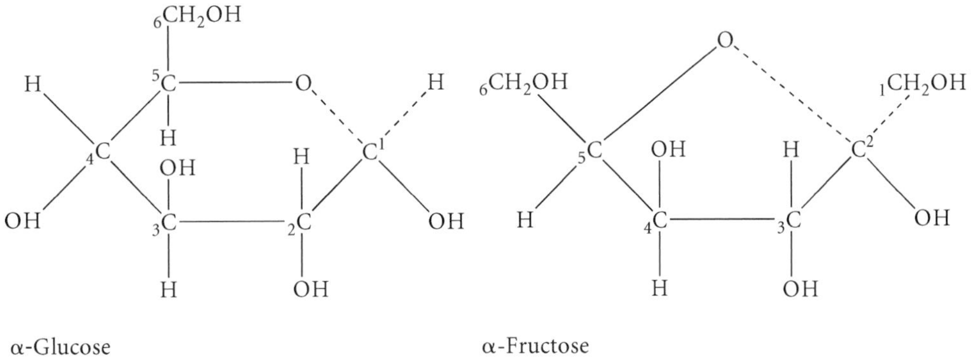

α-Glucose

α-Fructose

Figure 2.2 Structural formulae

(d) Look at Figure 2.2. What does this show? What is the difference between these two molecules? **[3]**

(e) Sucrose can be broken down into its two constituent monosaccharides by hydrolysis.
- **(i)** Write the word equation and the chemical equation for this reaction. **[2]**
- **(ii)** Draw a labelled molecular diagram to explain the chemical reaction. **[3]**

(f) Match carbohydrate compounds from the following list with the statements below. (Some statements require more than one answer.)
A glycogen, **B** sucrose, **C** glucose, **D** starch, **E** lactose, **F** fructose, **G** maltose, **H** galactose, **J** cellulose
- **(i)** storage carbohydrate found in liver and muscle tissue
- **(ii)** cane sugar; found also in sugar beet
- **(iii)** hexose monosaccharides
- **(iv)** the main carbohydrate in milk
- **(v)** contain β-glycosidic bonds
- **(vi)** have the chemical formula $C_{12}H_{22}O_{11}$
- **(vii)** structural carbohydrate found in plant cells
- **(viii)** condenses with glucose to form lactose
- **(ix)** carbohydrates that are chemically inactive
- **(x)** reducing sugars. **[18]**

2.5

Benedict's test is used to confirm the presence of a reducing sugar. The solution being investigated is added to blue Benedict's solution and the mixture heated in a water bath. The presence of reducing sugar is confirmed by a colour change to yellow/orange and the formation of a brick red precipitate.

The colour change is due to the reduction of blue copper ions Cu^{2+} in the Benedict's solution to Cu^{+} ions (brick red) in the precipitate. Reduction is brought about by an aldehyde group (–CHO) or by a keto group (C=O) in the molecular structure of the reducing sugar.

Look at the structural formulae of the following monosaccharides.

Glucose	Fructose	Galactose
^{1}CHO	$^{1}CH_2OH$	^{1}CHO
$H—{}^{2}C—OH$	$^{2}C=O$	$H—{}^{2}C—OH$
$HO—{}^{3}C—H$	$HO—{}^{3}C—H$	$HO—{}^{3}C—H$
$H—{}^{4}C—OH$	$HO—{}^{4}C—H$	$HO—{}^{4}C—H$
$H—{}^{5}C—OH$	$H—{}^{5}C—OH$	$H—{}^{5}C—OH$
$^{6}CH_2OH$	$^{6}CH_2OH$	$^{6}CH_2OH$

(a) Which of these is an aldehyde (or aldose) sugar? **[1]**

(b) Which is a keto (ketose) sugar? **[1]**

(c) Which will show a brick red precipitate on heating with Benedict's solution? **[1]**

(d) Only some disaccharides are reducing sugars. These have molecules composed of two monosaccharides that have condensed, leaving a free reducing group in the resulting disaccharide.
In reactions where the glycosidic bond forms between the aldehyde group of one monosaccharide and the keto group of the second, the resulting disaccharide is a non-reducing sugar.

Using the information given above predict the result of heating the following disaccharides with Benedict's solution. Which, if any, are reducing sugars?

Disaccharide	Monosaccharide units	Glycosidic bond
lactose	galactose, glucose	1–4
sucrose	glucose, fructose	1–2
maltose	glucose, glucose	1–4

[5]

(e) Disaccharide X is a non-reducing sugar. It was divided into two equal volumes. Solution 1 was heated with Benedict's solution. A few drops of dilute hydrochloric acid were added to solution 2 and it was heated with Benedict's solution. A brick red precipitate formed.

(i) What colour would be obtained when solution 1 was heated with Benedict's solution? **[1]**

(ii) Suggest an explanation for the result of the test with solution 2. **[2]**

2.6 Copy and complete the following table comparing four carbohydrates.

	Cellulose	Glycogen	Sucrose	Starch
Monomer			α-glucose	
Number of subunits		many		
Are side chains present?				yes
Type of bond(s) between monomers		α 1-4 glycosidic bonds		
Is it soluble in water?	no			
Where does it occur naturally in living cells?			in sugar beet and sugar cane; transported through plants in phloem tissue	
Function				storage carbohydrate

[7]

2.7 Proteins

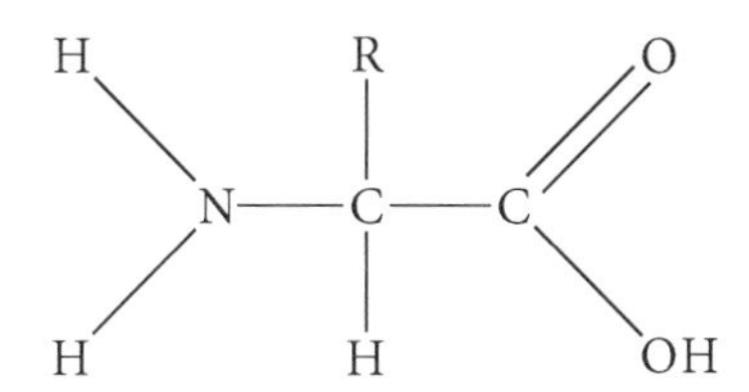

Figure 2.3 Structure of an amino acid

(a) Copy the molecule in Figure 2.3 adding labels for the amino group and the carboxyl group. What is 'R'? **[3]**

(b) Draw and fully label a diagram to show how two amino acid molecules react to form a dipeptide. Include a note of what type of reaction this is. **[5]**

(c) What name is given to a chain of amino acids? **[1]**

(d) What is the difference between an essential and a non-essential amino acid in the human diet? **[2]**

(e) Using words from the list below, complete the following passage.

amphoteric; isoelectric point; gain; lose; pH; negatively; basic; alkaline; dipolar

Buffers are used in biological systems to resist changes in ______________ when small volumes of acid or alkali are added. Amino acids are excellent buffers because their molecules have both ______________ and acidic characteristics: we say they are ______________________ .

In a neutral solution the amino acid is a ________________ion (or zwitterion) having both + and – charges. In acidic conditions amino acids ______________ hydrogen ions (H^+) from the medium and become positively charged. In ______________ solution amino acids ______________ electrons and so become ________________ charged. The pH at which an amino acid is electrically neutral is its __________________________. **[9]**

(f) Name one tissue in the human body where buffers are vital. **[1]**

(g) Approximately how many different amino acids are involved in forming polypeptides in human cells? **[1]**

2.8 The following terms describe protein structure

A Primary structure
B Secondary structure
C Tertiary structure
D Quaternary structure

(a) Match **A B C** or **D** to the correct statement **(i)–(iv)** :

(i) A protein consisting of more than one polypeptide chain.
(ii) A polypeptide chain coiled into an α-helix or folded into a β-pleat.
(iii) A polypeptide chain folded irregularly and held in place by hydrogen, ionic or sulphide bonds.
(iv) A sequence of amino acids forming a polypeptide chain. **[4]**

(b) Draw a series of labelled diagrams to show **A B C** and **D**.

(c) Proteins may be globular or fibrous.

Sort the following phrases into two lists, one relating to the characterisics of globular proteins and the second to fibrous proteins:

soluble in water, but usually form colloids because the molecules are large
little or no tertiary structure
tough
polypeptide chain folded into a tight spherical shape
insoluble in water
polypeptide chains with cross links
high tensile strength
complex tertiary structure **[8]**

(d) Which of the following proteins are globular proteins, which are fibrous proteins: amylase; keratin; albumen; insulin; tubulin; haemoglobin; collagen? **[7]**

2.9 Lipids

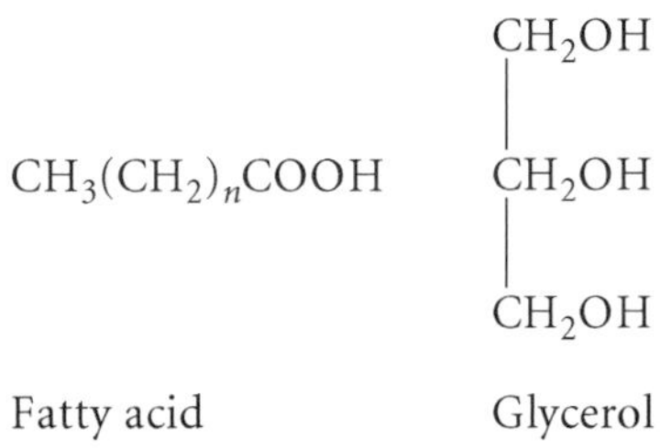

$$CH_3(CH_2)_nCOOH$$

Fatty acid

$$CH_2OH - CH_2OH - CH_2OH$$

Glycerol

Figure 2.4

(a) Using the information in Figure 2.4 draw and fully label a triglyceride. **[4]**

(b) What type of reaction occurs when a triglyceride is formed? **[1]**

(c) Explain the difference between a saturated fatty acid and an unsaturated fatty acid. **[1]**

(d) Complete the following passage:

Fats and oils are both ___________ . Both are synthesised from ___________ ___________ and ___________ . The difference between them is that fats are ___________ at 20°C whereas oils are ___________ . All lipids are insoluble in___________ but dissolve in ___________ ___________ such as ether or ___________ . Animals and plants contain waxy compounds synthesised from fatty acids and ___________ which often provide a waterproof layer as found for example in human ___________ . **[10]**

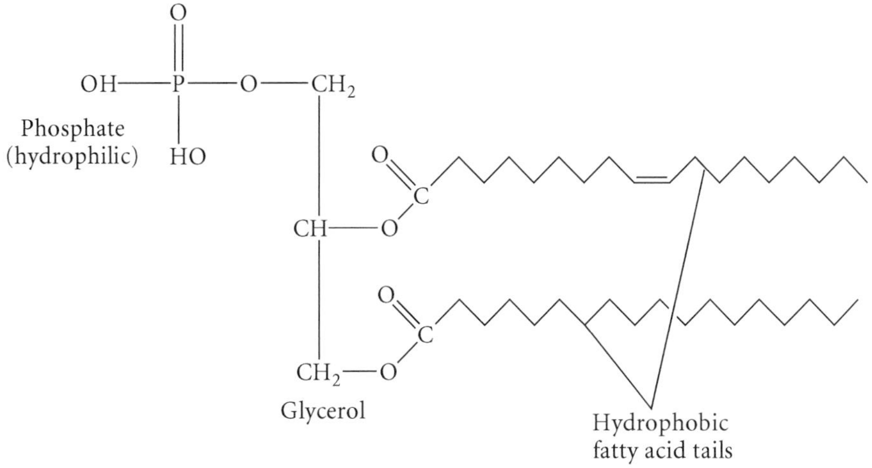

Figure 2.5 The general structure of a phospholipid

(e) How does the structure of a molecule of phospholipid differ from a triglyceride molecule? **[2]**

(f) Explain the labels 'hydrophilic' and 'hydrophobic'. **[2]**

(g) State two functions of phospholipids in humans. **[2]**

(h) Cholesterol is a steroid lipid (sterol).
Where is it synthesised in the human body? **[1]**
Name two hormones found in humans that are synthesised from cholesterol. **[2]**

2.10 Write a short account of the importance of lipids in animals. **[10]**

3 Enzymes

Enzymes are globular proteins that play a vital role catalysing metabolic reactions in all cells by reducing the activation energy needed for the reaction. Each is specific in its action and functions most efficiently at an optimum temperature and pH. Some require a cofactor in order to work. Enzymes do not get used up and therefore can be used many times. Enzyme action can be prevented by inhibitors, whose effect may be reversible or irreversible. Genetic mutations may have the effect of preventing the synthesis of a vital enzyme. Enzymes are of commercial importance and are often immobilised when used in industrial processes to increase their stability.

3.1 **(a)** Fill the gaps with appropriate words to complete the passage about enzymes.

Enzymes are found in all living ____________ . Chemically they are ____________ and have a ____________ structure. They are catalysts because they ____________ chemical reactions. As with inorganic catalysts, enzymes are not ____________ nor____________ ____________ during the reactions they catalyse. Enzymes are ____________ in their action, many catalysing a single reaction only. Enzymes are most efficient at ____________ levels of pH and temperature. Most enzymes are ____________ at temperatures above 60°C. Their action may be reduced or stopped by chemicals known generally as ____________ . **[10]**

(b) Copy and complete the following table of information about some common enzymes.

Enzyme	**Metabolic process involved in**	**Function**
decarboxylase	Krebs cycle	
		hydrolysis
DNA ligase		
	electron transfer chain	
		removal of hydrogen

[9]

3.2 The energy needed to make or to break molecular bonds to start a chemical reaction, is called activation energy. Look at Figure 3.1

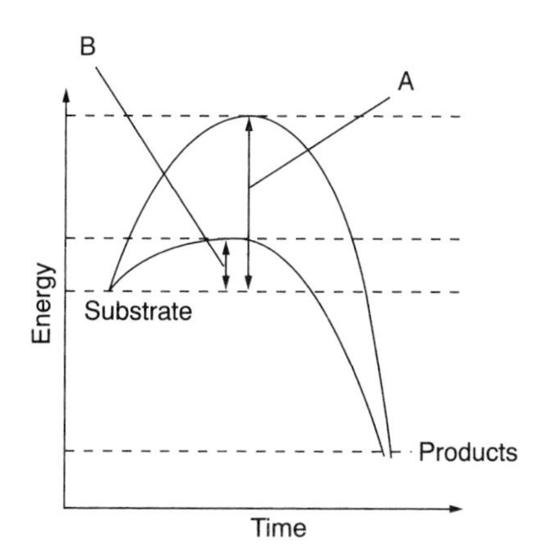

Figure 3.1

(a) Suggest a way to raise the activation energy of reactants **without** using an enzyme or an inorganic catalyst. **[1]**

(b) What does A represent? **[1]**

(c) What is B? **[1]**

(d) Give two examples of enzymes that are specific for a single substrate and the reactions that take place. **[4]**

(e) Name two enzymes that can catalyse more than one reaction and explain what reactions are involved. **[4]**

(f) How are enzymes vital to the control of the metabolic reactions in a cell? **[1]**

(g) What controls which enzymes will occur in any one cell? **[1]**

3.3 **(a)** Write a short account to explain the difference between the 'lock and key hypothesis' and the 'hypothesis of induced fit' to explain enzyme action. You should include the following terms in your account:

substrate; products; active site; amino acids; conformation; enzyme-substrate complex

[8]

(b) What scientific technique might be used to provide evidence for the hypothesis of induced fit? **[1]**

3.4 **(a)** Explain the term enzyme inhibitor. **[1]**

(b) What information about enzyme structure and enzyme action might be gained from the study of enzyme inhibitors? **[2]**

(c) Read the two statements below and name the term that each one describes:

(i) Two substrates having very similar molecular structure so that either may bind to the active site of a specific enzyme. This is likely to result in a slowing down of the enzyme-catalysed reaction.

(ii) A molecule that combines permanently with an enzyme at its active site or at another site on its tertiary structure. This distorts the enzyme's active site stopping the enzyme-controlled reaction.

Statement **(i)** is ______________________________ **[1]**

Statement **(ii)** is ______________________________ **[1]**

(d) Give a named example of a competitive enzyme inhibitor and explain briefly how it might act in a cell. **[3]**

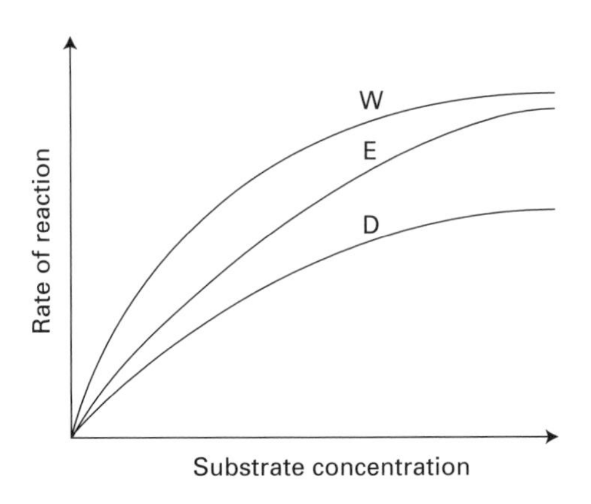

Figure 3.2

(e) Look at Figure 3.2.
The graph shows the results of an investigation measuring the rates of enzyme-controlled reactions. Line W shows the data from the control test. A second experiment was carried out under identical conditions except that a competitive inhibitor was added to the enzyme-substrate mixture. In a third experiment a non-competitive inhibitor was added to the enzyme-substrate mixture.

(i) Explain in principle how the first test was carried out. **[4]**

(ii) Which line shows the results from the second experiment? **[1]**

(iii) Explain your answer to question **(ii)**. **[4]**

(f) What is a cofactor? Why are cofactors important in some enzyme controlled reactions? **[2]**

3.5

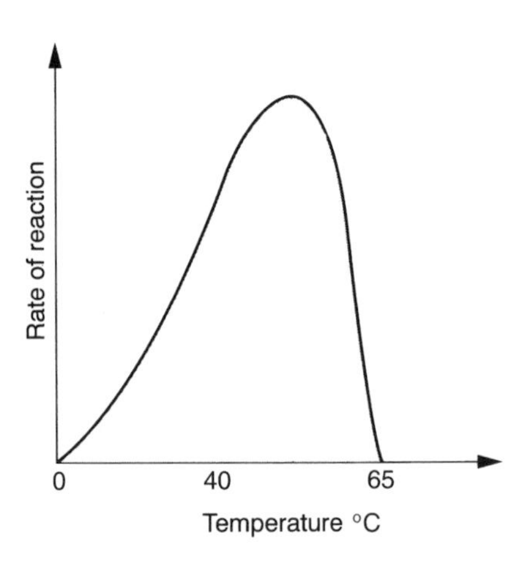

Figure 3.3

(a) Using the graph given in Figure 3.3, give an account of the effects of temperature on enzyme activity. **[8]**

(b) Explain the term temperature coefficient (Q_{10}). **[3]**

3.6 Catalase is an enzyme found in many plant and animal cells. It breaks down hydrogen peroxide, a toxic by-product of many metabolic reactions, into water and oxygen. The apparatus shown in Figure 3.4 can be used to measure catalase activity at different pH conditions.

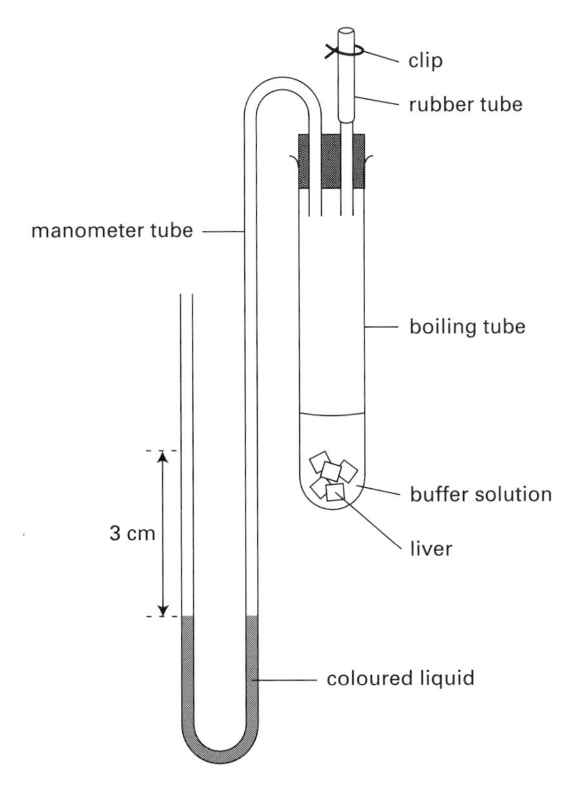

Figure 3.4

(a) Explain the term pH. **[2]**

(b) What is the purpose of the buffer solution in which the cellular material is suspended? **[1]**

(c) What variables must be kept constant during this experiment? **[3]**

A group of students used the apparatus to determine the optimum pH for catalase. Cubes of liver were placed in buffer solutions ranging from pH 5–9. Hydrogen peroxide solution was added to each boiling tube in turn and the time measured for the fluid in the manometer tube to rise by 3 cm. Each test was repeated twice. The students' raw results are given in the table below.

pH	Experiment 1	Experiment 2	Experiment 3	Average time
5	7 min 40 s	8 min 17 s	8 min 3 s	8 min
6	2 min 14 s	2 min 7 s	1 min 39 s	2 min
7	20.8 s	17.9 s	17.1 s	18.6 s
8	17.4 s	14.0 s	13.6 s	15 s
9	10 min 40 s	9 min 57 s	9 min 23 s	10 min

Rate of reaction was worked out using the formula: $\frac{100}{\text{average time in seconds}}$

(d) Calculate the rate of reaction for each pH. **[2]**

(e) Plot a graph of rate of reaction against pH. **[4]**

(f) When the students had drawn their graph, they decided to carry out one further test at pH 7.5. This showed average reaction time at pH 7.5 was 13.3 seconds. Why was this extra measurement made? **[1]**

(g) Amend your graph to include the rate for activity at pH 7.5 **[2]**

(h) What is the optimum pH of catalase as shown by these experimental results? **[1]**

(i) Explain why catalase activity was so slow at pH 9. **[2]**

(j) How would you modify this experiment to determine the optimum temperature for catalase activity? **[2]**

3.7 Look at Figures 3.5 and 3.6 which relate to a reaction in which enzyme X catalyses the hydrolysis of substrate Y. Temperature (30°C) and pH (7.0) were constant during both tests. In the first test enzyme concentration was constant and in the second substrate concentration remained constant.

(a) Compare the two graphs, explaining any differences you describe. **[4]**

(b) If temperature and pH remain constant, suggest one reason why the rate of reaction shown in Figure 3.6 might decrease rapidly. **[1]**

(c) What is a 'limiting factor'? Give three examples of limiting factors saying how they bring about an effect. **[7]**

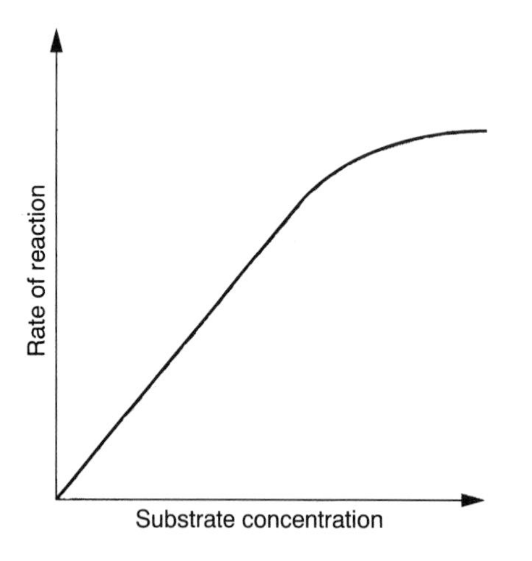

Figure 3.5

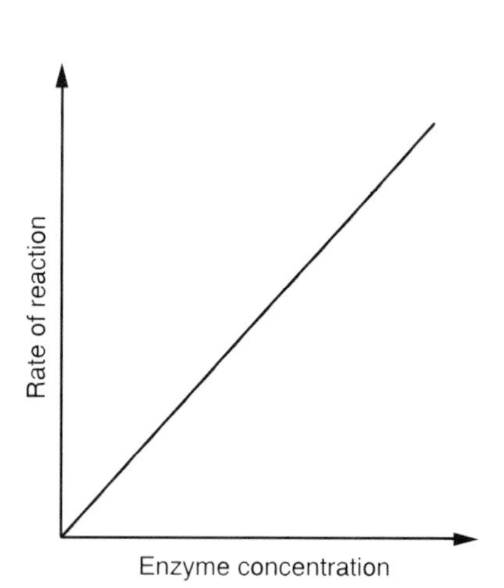

Figure 3.6

3.8 Applications of enzymes

Lactase is an enzyme which hydrolyses lactose, the sugar occurring in milk, into glucose and galactose. Around 75% of humans world wide are lactose intolerant, due to their inability to secrete lactase. The lowest incidence of the condition is amongst adults in north west Europe. Avoidance of dietary lactose controls the disorder.

In a laboratory experiment cow's milk was poured through a column of lactase immobilised in beads of calcium alginate. Treated milk was collected from the bottom of the syringe.

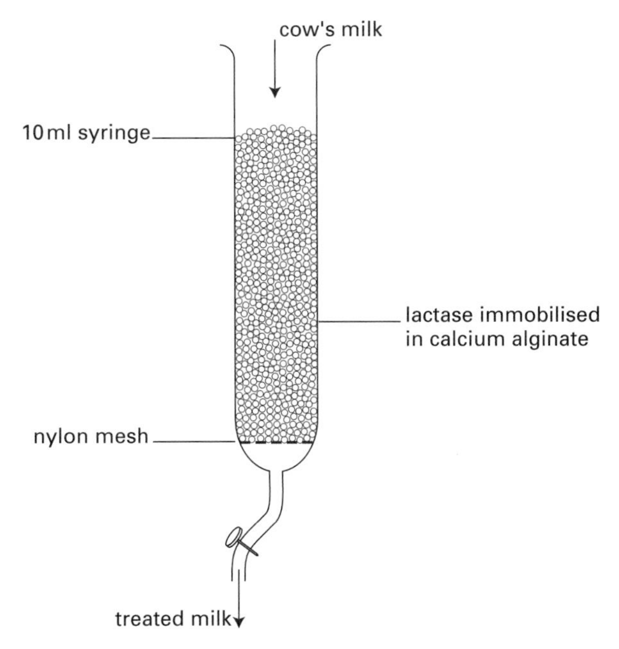

Figure 3.7

(a) Explain three methods of immobilising enzymes. **[3]**

(b) What are the advantages of using immobilised enzymes in biotechnological processes? **[3]**

(c) How would the untreated milk differ from the end product of this experiment? **[2]**

(d) Describe briefly two experimental tests you could perform to justify your answer to **(c)**. **[4]**

(e) Galactose is an inhibitor of lactase. Predict the effect that different flow rates of cow's milk in this experiment might have on the final results. **[4]**

(f) Why are most babies able to digest milk while 75% adult humans are lactose intolerant? **[2]**

Lactolite is a milk modified for lactose intolerant people. Provamel is a brand of soya milk. The following table gives the constituents of whole cow's milk, Lactolite and Provamel:

	Whole milk	Lactolite	Provamel
Energy	284 kJ/68 kcal	284 kJ/68 kcal	149 kJ/36 kcal
Protein	3.2 g	3.2 g	3.6 g
Carbohydrates	4.8 g	4.8 g	0.6 g
– of which sugars	4.8 g	4.8g	0.1 g
– lactose	4.7 g	less than 0.24g	not given
– glucose	NIL	2.4 g	not given
Fat	4.0 g	4.0 g	2.1 g
– of which saturates	2.5 g	2.5 g	0.4 g
Fibre	NIL	NIL	1.2 g
Calcium	119 mg	119 mg	120 mg

(g) Comment on the data given in the table above. **[3]**

(h) What dietary deficiencies might occur in humans who are lactose intolerant? **[1]**

(i) During the last 50 years the ethnic composition of the population of north west Europe has changed. Explain the implications this might have on the incidence of lactose intolerance in that population? **[3]**

(j) List two differences between human breast milk and whole cow's milk. **[2]**

3.9 *Fruits and berries used to make fruit juice contain varying amounts of pectin. It is responsible for the hardness of unripe fruit. As the fruit ripens the pectin is broken down into soluble products by pectolytic enzymes, and so the fruit becomes softer.*

In the production of apple juice, the apples are first chopped and crushed. Pectin may then pass out of the apple pulp, increasing the viscosity of the juice and making it difficult to obtain the maximum yield. The pressed apples are filtered and the juice collected, but it may be cloudy, poor in colour and in flavour. These problems can be overcome by treating the crushed apples with the enzyme pectinase, adding more to the juice after filtration.

(a) Using the information in the passage draw a flow chart to show the stages involved in the commercial production of apple juice. **[5]**

(b) Where in an apple is pectin found? **[1]**

(c) What happens when pectinase is added to pectin? **[2]**

(d) Why are the apples chopped and crushed? **[1]**

(e) Suggest three reasons for the addition of pectinase during the manufacture of apple juice. **[3]**

(f) Suggest another enzyme which might be added during production, to increase the yield of apple juice. Explain your choice. **[2]**

(g) If you were starting as a commercial producer of apple juice, what factors would you consider when deciding which type of apple to use? **[3]**

4 Breathing and gas exchange

Respiration is a chemical process that transfers energy from food to the high energy molecule ATP which acts as an energy store for other chemical processes happening in cells. Human cells respire aerobically most of the time, so oxygen from air is a vital raw material. This is gained by the mechanical process of breathing and subsequent gaseous exchange at the alveolar surfaces.

4.1 Lung structure

Figure 4.1 shows diagrams of epithelial tissues found in the human respiratory system.

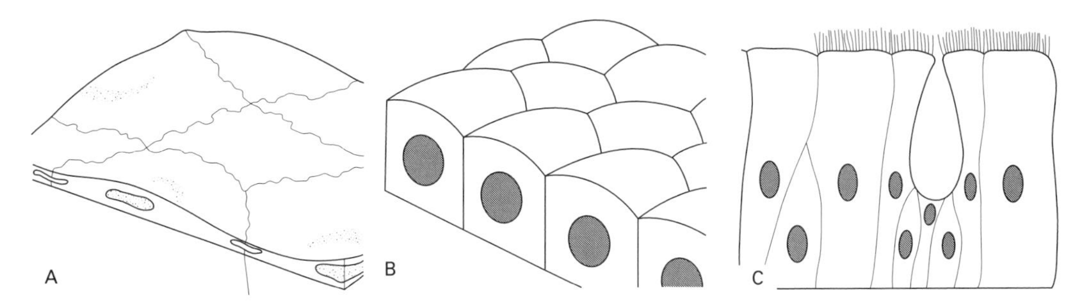

Figure 4.1

(a) Which of the epithelia shown is present in the bronchioles? **[1]**

(b) Which other tissue mainly makes up the walls of bronchioles? **[1]**

(c) Which of these tissues lines the trachea? **[1]**

(d) How is the structure of the epithelium named in **(c)** linked to its position and function? **[3]**

(e) Which of these epithelia is most likely to allow diffusion of gases at the fastest rate? **[1]**

(f) Describe the properties that an efficient gas exchange surface would have. **[3]**

4.2 Read this passage about the structure of the lungs and complete it by filling in the blanks with appropriate terms.

The composition of inhaled air alters as it passes on its way to the lungs as it is __________ and __________ is added to it. The __________ divides into two bronchi, one bronchus passing to each lung. Cartilage rings in the trachea give it sufficient strength to remain open, even when the __________ __________ inside it drops. All but the smallest __________ also contain some cartilage for this reason. Alveoli are the blind ended sacs which hugely increase the ________ ________ of the gas exchange surface within the lungs. The epithelial cells within the alveoli produce a phospholipid which acts as a __________, helping to prevent collapse of these tiny sacs that could be caused by the high _________ _________ between the wet surfaces. **[8]**

4.3 Gaseous exchange

Figure 4.2 represents gas diffusion across the alveolar surface.

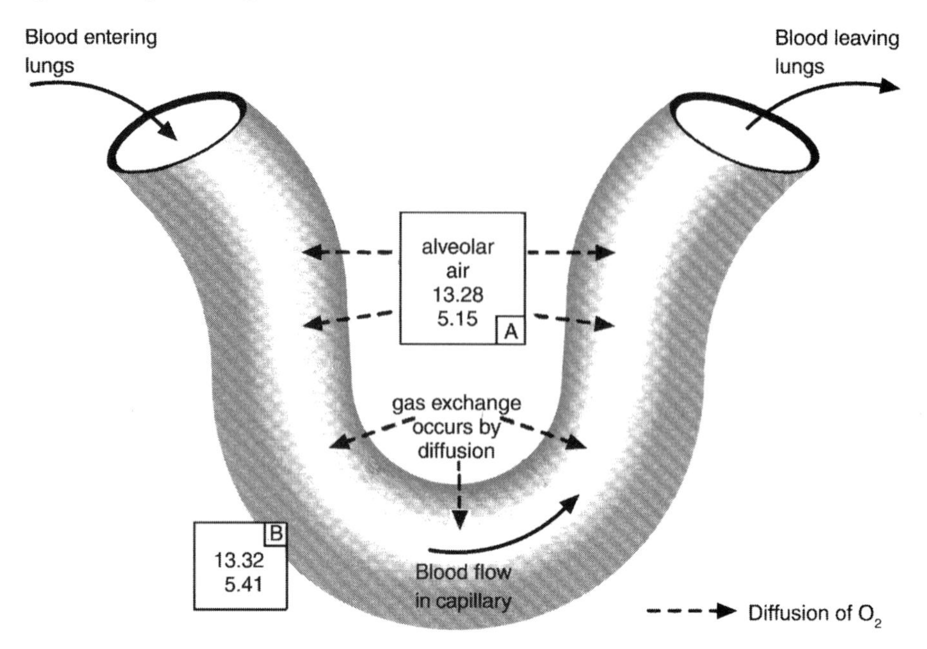

Figure 4.2

The partial pressure of a gas is the pressure it exerts as part of the total gas pressure exerted by a mixture of gases. This unit of pressure can be used to interpret the direction of gas diffusion. Several partial pressures for oxygen are shown in the diagram.

(a) Select two figures for partial pressure of oxygen (one from box A and one from box B) which might occur when oxygen is diffusing into the blood capillary from the alveolus, and add the correct unit.

_____________ from box A

_____________ from box B. **[2]**

(b) Explain why these partial pressures cause oxygen to diffuse into the blood capillary but not out of it. **[2]**

(c) Would you expect the partial pressure of oxygen in tissue cells to be higher or lower than that of oxygen in the blood capillaries around the alveoli? **[1]**

(d) Why is there a difference between the partial pressure of carbon dioxide in a tissue cell and in the blood in a capillary located around an alveolus? **[3]**

4.4 Fick's law states that the rate of diffusion (through a membrane) is proportional to:

$$\frac{\text{surface area} \times \text{difference in concentration}}{\text{thickness of membrane}}$$

(a) Explain the relevance of each of the components of this equation. **[3]**

(b) Suggest three other factors which would influence the speed of diffusion of a molecule. **[3]**

4.5 Ventilation

Write an account of how ventilation movements (breathing) bring about mass flow of air in and out of the lungs. **[8]**

4.6 Figure 4.3 represents the lungs during inspiration.

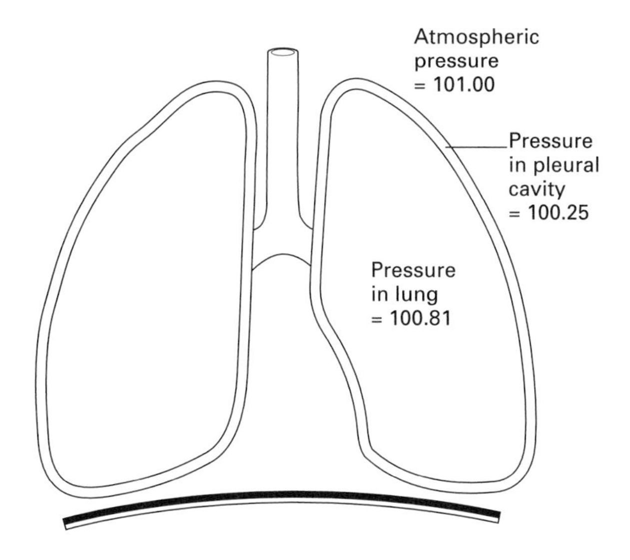

Figure 4.3

(a) What units are used for the pressures shown as numbers in the diagram? **[1]**

(b) From the data given, how can you tell that this diagram represents inhalation? **[1]**

(c) What is elastic recoil and how does it play a part in breathing? **[1]**

(d) Why does a hole which punctures the pleural membranes and wall of the thorax interfere with breathing? **[3]**

4.7 **(a)** Distinguish between the terms tidal volume and ventilation rate. **[2]**

(b) The tidal volume for humans at rest is about 450 cm^3. A person breathes 102 times during a 10 minute radio programme. Calculate their ventilation rate, showing your workings clearly. **[3]**

(c) What might cause the tidal volume to increase over a period of time? **[1]**

(d) What is meant by residual volume? **[1]**

4.8 A spirometer is a machine for measuring the amount of air moving into and out of the lungs as we breathe. This provides information for medical staff about the health condition of a person's lungs. Figure 4.4 shows a spirometer trace.

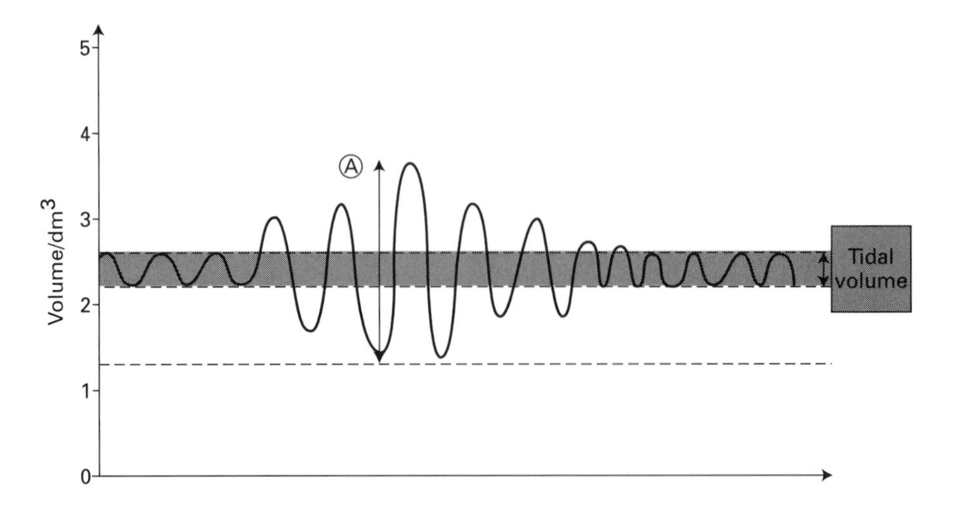

Figure 4.4

(a) What variable is measured along the X-axis of the spirometer trace? **[1]**

(b) What is tidal volume? **[2]**

(c) The spirometer trace has changed during the period marked at A, because the person begins to exercise. What is the term used to describe this increased volume? **[1]**

(d) What is meant by residual volume? **[1]**

(e) About what volume (dm^3) is represented by the residual volume in human lungs? **[1]**

4.9 Diffusion and surface area

Two cubes are shown in Figure 4.5.

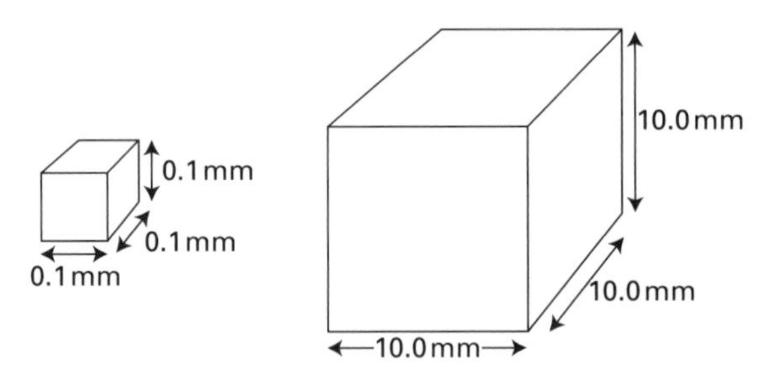

Figure 4.5

(a) Work out the surface area and the volume for each cube. **[4]**

(b) Calculate the surface area to volume ratio for each cube. **[2]**

(c) Imagine that these cubes represent two different animals. Which animal is unlikely to be able to obtain sufficient oxygen by diffusion alone? **[1]**

(d) Why do some animals require mass flow of air in and out of the body? **[2]**

4.10 Asthma

Asthma is a common respiratory condition, particularly in children.

(a) Describe physical changes in the airways that happen during an asthma attack. **[2]**

(b) What are the symptoms of asthma? **[1]**

(c) Name environmental factors which can trigger an asthma attack. **[2]**

(d) Describe the elements of a self help plan for managing asthma. **[3]**

4.11 A nebuliser can be used by someone who has asthma to help relieve an extremely difficult attack. Essentially it is an air pump which carries a fine mist of the appropriate medicine into the airways over a period of a few minutes. Figure 4.6 shows the breathing pattern for a person who has asthma, over a period of 5 days using a nebuliser.

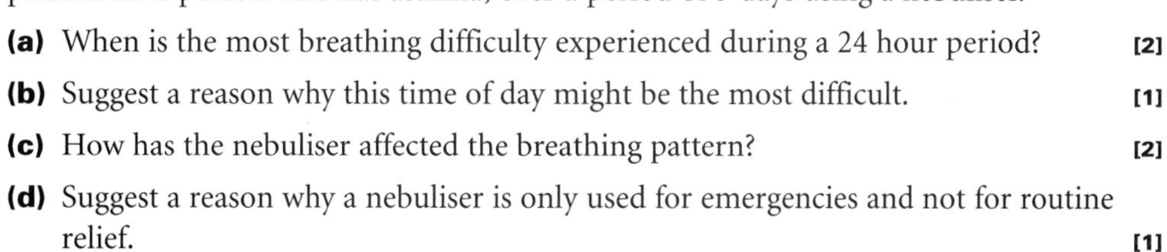

(a) When is the most breathing difficulty experienced during a 24 hour period? **[2]**

(b) Suggest a reason why this time of day might be the most difficult. **[1]**

(c) How has the nebuliser affected the breathing pattern? **[2]**

(d) Suggest a reason why a nebuliser is only used for emergencies and not for routine relief. **[1]**

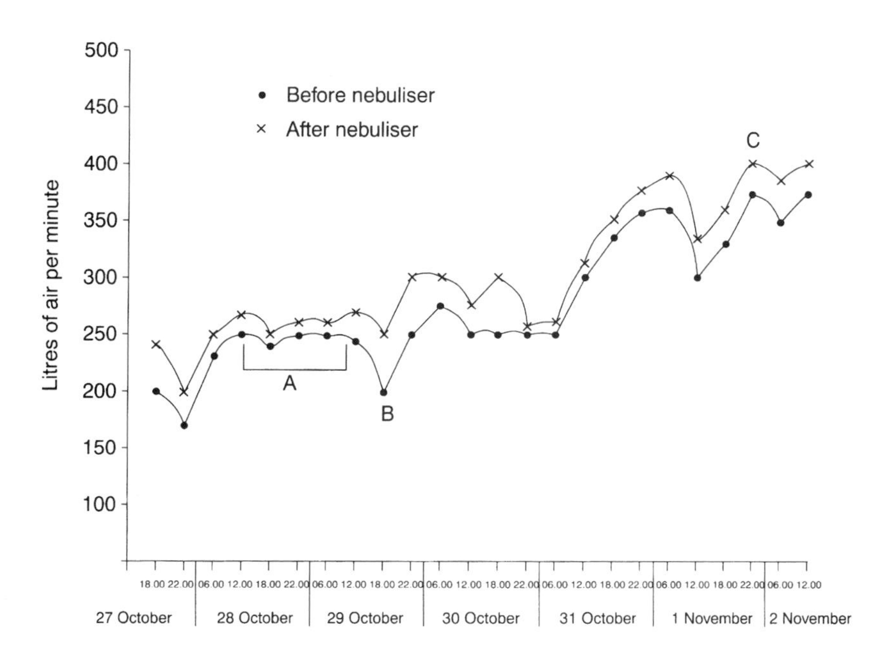

Figure 4.6

4.12 Smoking and health

Around 15–20% of all deaths in the UK are related to tobacco smoking. Below is a table containing information about the health hazards associated with smoking.

Condition	Effects	Cause
Cardiovascular disease	more cholesterol deposits in blood and increased risk of heart attack	increases rate of low density lipid formation and increases clots
1	low body mass	reduction in placental circulation and oxygen carriage
Cancer	tumours	tars/other carcinogens
Emphysema	lungs fill with mucus; breathlessness	2
Poor circulation	Constriction of arterioles	nicotine constricts smooth muscle

(a) Write phrases which would be appropriate to fill the blanks numbered **1** and **2** in the table. **[2]**

(b) Which of the substances named in the table is a stimulant? **[1]**

(c) Which substance in tobacco smoke reduces the uptake of oxygen by haemoglobin? **[1]**

(d) What is a carcinogen? **[1]**

(e) Why and how do substances in tobacco smoke inhaled by a pregnant woman affect the developing fetus? **[3]**

(f) Suggest a reason why a low body mass at birth might be a disadvantage to babies. **[2]**

4.13 Figures 4.7 to 4.9 show trends in smoking and the effects of smoking on mortality.

(a) Describe the trend in smoking for women between 1950 and 1970. **[3]**

(b) Suggest reasons why the trends are different for men and women. **[3]**

(c) Compare graphs 4.7 and 4.8 for females. **[3]**

(d) Comment on the trends in Figure 4.9. **[3]**

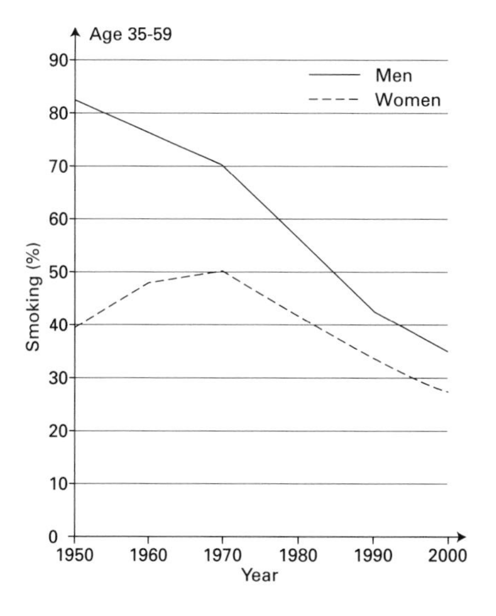

Figure 4.7 Trends in smoking in the UK, for the age group 35–59

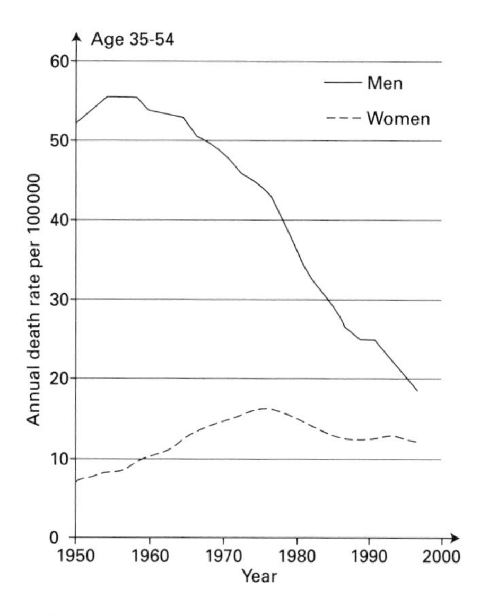

Figure 4.8 Trends in mortality from lung cancer in the UK in the age group 35–54

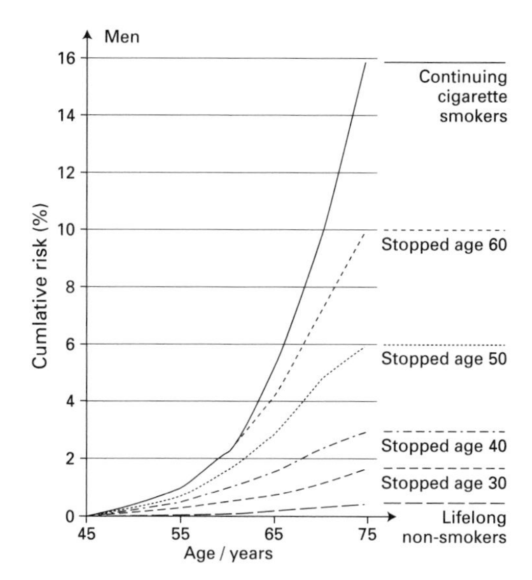

Figure 4.9 Effects of stopping smoking on the cumulative risk (%) of death from lung cancer in UK males in 1990

5 Blood and circulation

The cardiovascular system comprises the heart, blood vessels and blood. Its main role is distribution, vital in maintaining conditions within body tissues. The blood carries materials by mass flow between cells and exchange surfaces, such as those in the alveoli, intestines and glands. Closely linked with this system is the lymphatic system. There is a movement of fluid from plasma to the tissues, from the tissues into lymphatic vessels, and finally back into the blood circulation.

5.1 Structure of blood

Copy and complete the table of information concerning the composition of blood.

Component	Feature	Function
Plasma	90% water containing electrolytes such as __________ (1) metabolites, eg __________ (2) and proteins such as __________ (3)	__________ (4) __________ (5)
White blood cells eg __________ (6) and __________ (7)	__________ (8) __________ (9)	mainly defence against disease and immunity
Red blood cells (erythrocytes)	__________ (10) __________ (11)	__________ (12) __________ (13)
Platelets	these are formed from __________ (14) and they are __________ (15) numerous than red or white blood cells	__________ (16)

[16]

5.2 Haemoglobin

The table below shows data about haemoglobin (Hb) levels in a group of new blood donors.

Blood is not normally taken from a donor if the Hb levels fall below 135 g/l in males and 125 g/l in females. The table gives a comparison of donor Hb (g/l), and percentage of donors below UK Blood Transfusion Service (BTS) recommendations for acceptance.

	Mean Hb(g/l)	SD	% donors with less than g/l Hb shown				
			135	125	115	105	95
Male	144.0	8.2	13.1	0	0	0	0
Female	142.0	8.5	–	2.6	0.2	0	0

(a) Why is the Hb level of prospective blood donors monitored in the UK? **[1]**

(b) The column headed SD in the table refers to the standard deviation test. What is the standard deviation test, and what is the value of using it? **[2]**

(c) Suggest a reason why no males in this sample have a Hb level of less than 125 g/l. **[1]**

(d) What percentage of women had less than 125 g/l Hb? **[1]**

(e) What percentage of men had more than 135 g/l Hb? **[1]**

5.3

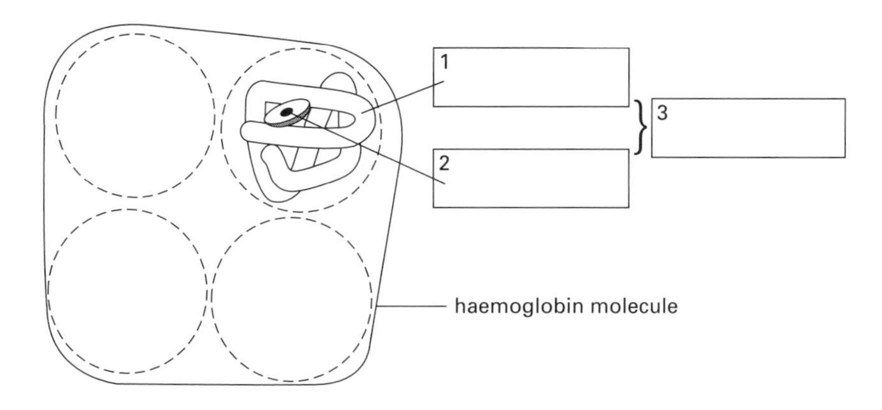

Figure 5.1 Diagram of a molecule of haemoglobin

(a) Suggest appropriate terms for the labels left blank on Figure 5.1. **[3]**

(b) One red blood cell may contain 250 molecules of haemoglobin. How many oxygen atoms could this cell carry if it was completely saturated with oxygen? **[2]**

(c) Why is very little oxygen carried by plasma? **[1]**

(d) What is the name of the compound formed when carbon dioxide combines with haemoglobin? **[1]**

(e) Which blood condition is caused by a lack of iron? **[1]**

5.4 Figure 5.2 is a graph showing the oxygen dissociation curve for human haemoglobin.

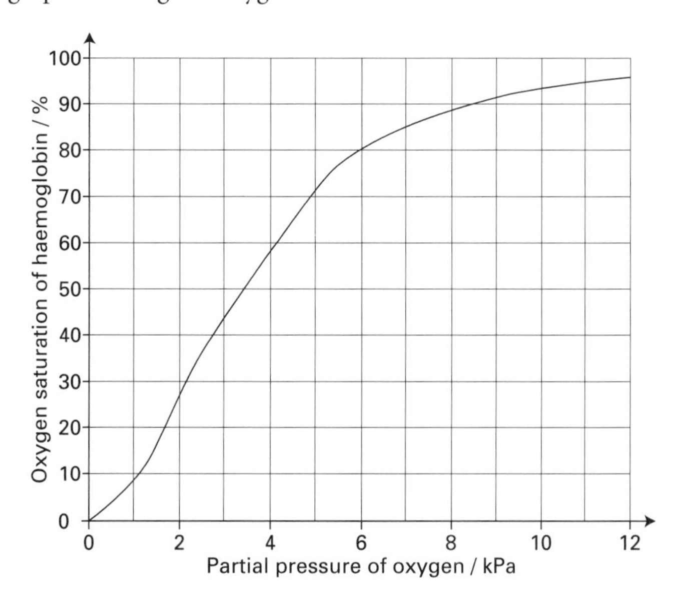

(a) Calculate the difference between the loading tension and the unloading tension for this sample of haemoglobin. **[2]**

(b) Explain briefly why the dissociation curve is S shaped. **[2]**

(c) Copy the diagram and sketch in the curve which might occur if the partial pressure of carbon dioxide increased. **[1]**

(d) Why is the Bohr shift important to cellular function? **[2]**

5.5 Circulatory disorders

The table below refers to blood and circulatory disorders. If a feature is correct for a disorder put a tick in the appropriate box and if it is not correct place a cross in the box.

	Feature			
Disorder	**Is a genetic disorder**	**Iron content is low**	**Low blood sugar occurs**	**Oxygen carriage is reduced**
Anaemia				
Haemophilia				
Sickle cell				
Hyperglycaemia				

[4]

5.6 Structure of vascular system

Write an account summarising the main structural features of arteries, capillaries and veins, relating the features to the functioning of each type of vessel. **[12]**

5.7 Below is a list of descriptions. Match each item on the list with one of the three blood vessel types: capillary, artery, vein.

A accounts for over 60% of systemic circulation
B low blood pressure (around 1kPa)
C speed of blood flow is fastest
D tunica media is thin
E site of exchange between blood and tissues
F semilunar valves present
G carry blood to the organs
H pulsatile blood flow **[8]**

5.8 **(a)** How is the speed of blood flow related to blood pressure and cross-sectional area of a blood vessel? **[2]**

(b) Suggest reasons why rate of blood flow varies in different parts of the circulatory system. **[2]**

5.9 Figure 5.3 represents the heart structure.

(a) Annotate the heart as fully as possible. **[25]**

(b) Where is the human heart located? **[2]**

(c) What is the function of pericardial fluid? **[1]**

(d) What is the function of the coronary circulation? **[1]**

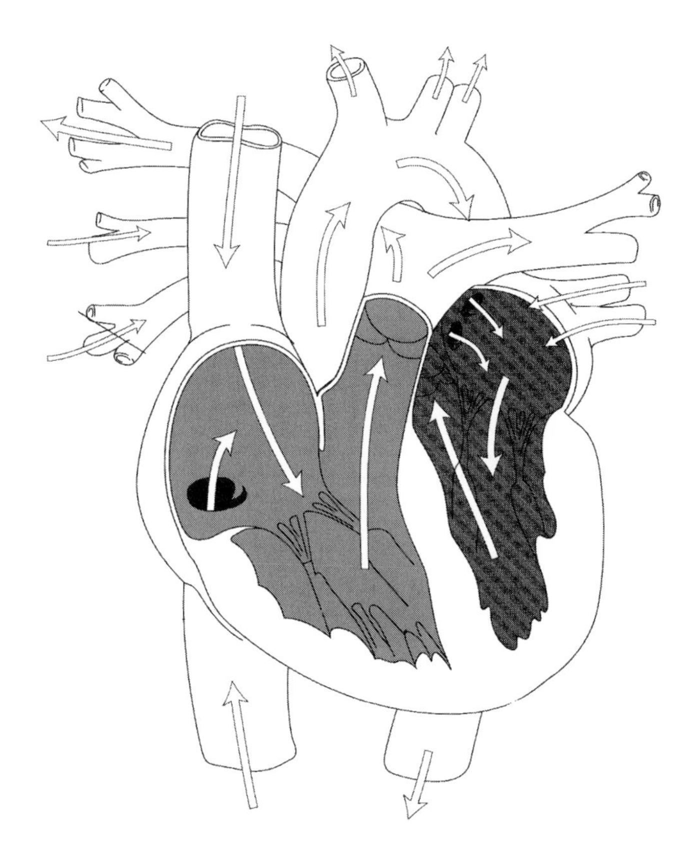

Figure 5.3

5.10 Copy the following passage, filling in the blank spaces with appropriate words or phrases.

Cardiac muscle tissue has unique features adapting it for its special role as a __________. The muscle __________ which make up cardiac muscle tissue are __________ by cross-bridges. This integrates the tissue and allows contraction in all __________. The ______________ ____________________ help to keep the cells in position as contraction is occurring. Since __________ occurs rapidly in this tissue at all times, many more mitochondria are present than in other muscle types. External stimulation is not needed to bring about contraction, and because of this cardiac muscle is described as __________. **[7]**

5.11 Heart beat

Figures 5.4 A–D represent stages in the heart cycle.

(a) Place these diagrams in the correct sequence, starting with diagram 5.4A. **[2]**

(b) In which diagrams is the blood pressure increasing in the atria? **[1]**

(c) What causes the bicuspid and tricuspid valves to close? **[1]**

(d) Why is it necessary for the bicuspid and tricuspid valves to close during the heart cycle? **[1]**

(e) What blood pressure differential exists between the main arteries and the ventricles as the semi-lunar valves begin to close? **[1]**

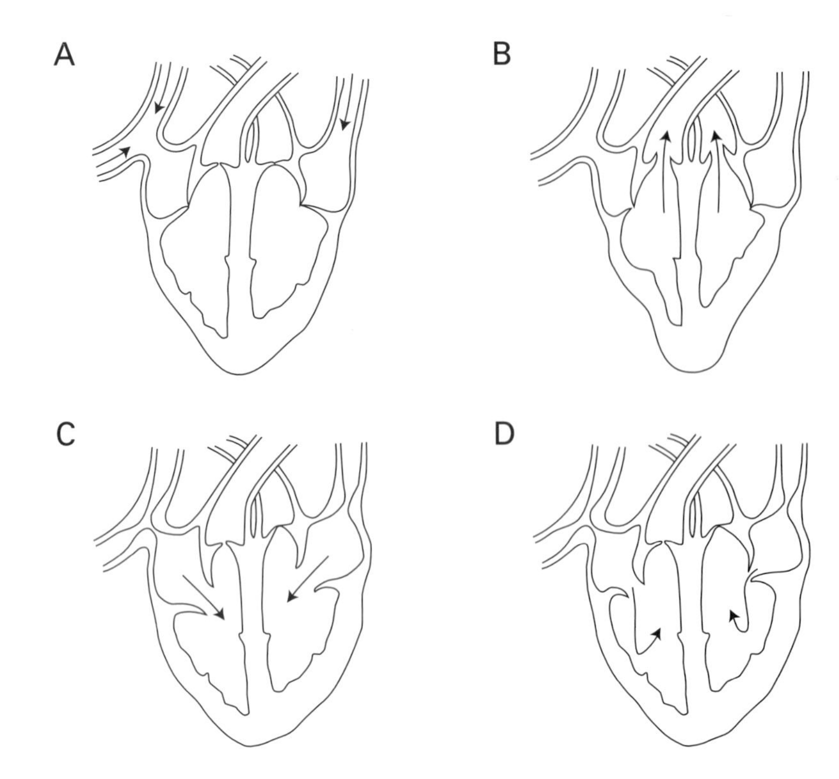

Figure 5.4

5.12

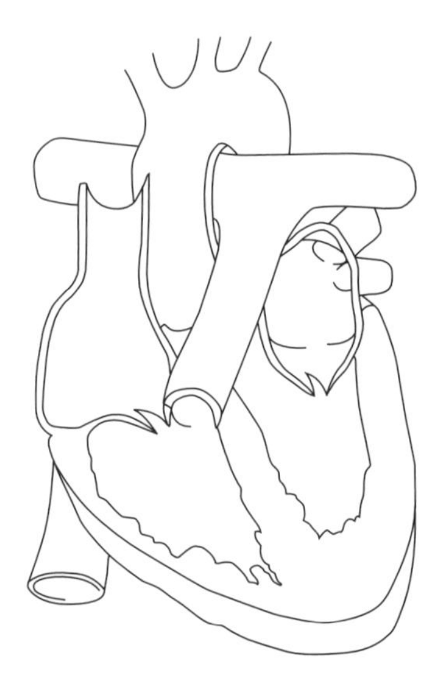

Figure 5.5 Outline of the human heart

(a) Copy the diagram and mark on it the sinu-atrial and atrio-ventricular nodes. **[1]**

(b) What would be the effect of the sinu-atrial node not functioning? **[4]**

(c) What other nervous tissue is important in the heart cycle? **[2]**

(d) What is the role of the AV node? **[2]**

(e) Sketch the position and label other important nerve tissue on the diagram, showing the direction that nerve impulses travel. **[3]**

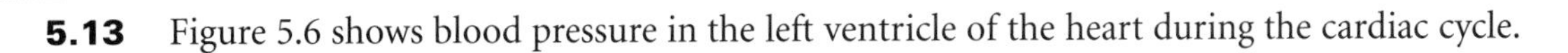

5.13 Figure 5.6 shows blood pressure in the left ventricle of the heart during the cardiac cycle.

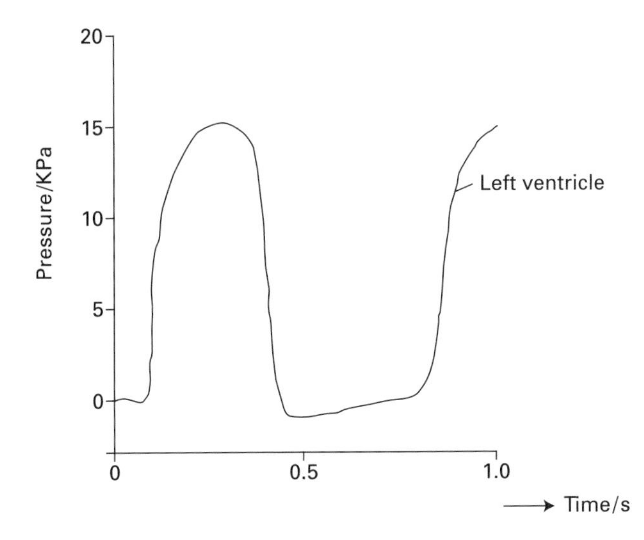

Figure 5.6

(a) Copy the diagram and sketch the shape of curve expected for the right side of the heart. **[2]**

(b) Explain why there are pressure differences between the two ventricles. **[2]**

(c) Mark on the curve for the left ventricle an X where ventricular diastole occurs. **[1]**

(d) What is meant by atrial systole? **[1]**

5.14 Figure 5.7 is taken from a display on a visual display unit (VDU) and shows the electrical activity of the heart as it beats.

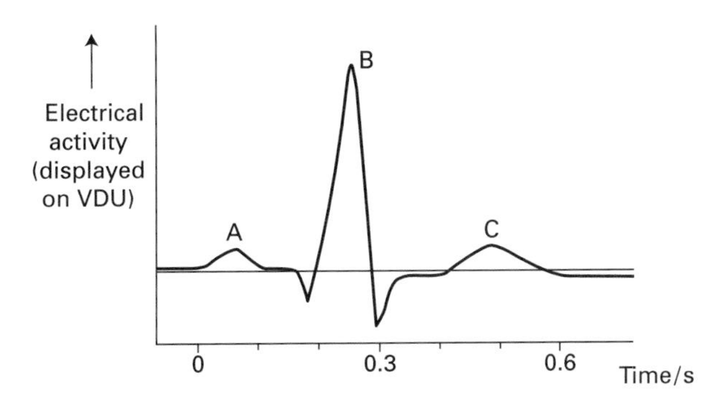

Figure 5.7

(a) What is this type of chart called? **[1]**

(b) How is the sinu-atrial node involved at the stage of the heart cycle marked A? **[2]**

(c) At point C the ventricles repolarise. What does this mean? **[2]**

5.15 Circulation

The blood circulation through the heart of a developing fetus is different from that of a child after birth.

(a) Why is there no direct connection between the maternal blood and the fetal blood? **[1]**

(b) How does exchange of materials between fetal and maternal blood occur, and why is it necessary? **[4]**

(c) What is the foramen ovale, and what is its role? **[2]**

(d) 90% of the blood flow through the fetal pulmonary artery is diverted to the fetal aorta. How does this occur, and for what reasons? **[3]**

(e) What changes happen to the heart circulation of a newly born infant? **[2]**

5.16 Anaemia

The IQ test is one form of intelligence test. A group of 139 girls from an English comprehensive school were tested. Some of them were showing iron deficiency anaemia (IDA); others were iron deficient but were not anaemic (ID) and others were not iron deficient (IR). The girls were tested for IQ before and after treatment. Some received a safe level of iron treatment and others received a placebo dose. The results are shown below:

	Girls given iron treatment			**Girls given placebo treatment**		
	IDA	ID	IR	IDA	ID	IR
First IQ	107	111	114	107	107	109
Second IQ	113	118	118	111	110	118
P value	< 0.001	<0.001	<0.001	0.173	0.150	<0.001

P value = a level of significance, where <0.001 means that the chance of these results not being significant is less than 0.001 or 0.01% i.e. these are assumed to be significant.

(a) What is the main role of iron in the body? **[1]**

(b) The IQ score for all participants increased between the first and second test. Suggest a reason unrelated to iron in the diet which might account for this. **[1]**

(c) What does this data suggest about the effect of using iron supplement? **[2]**

(d) What is a placebo treatment? **[1]**

(e) Suggest reasons why being iron deficient might affect mental performance. **[2]**

6 The body and exercise

Both immediate and longer term changes happen in the body because of exercise. As part of the body's homeostatic response, the immediate changes allow the body to cope with varying demands at the cellular level, for example for oxygen or glucose, helping to keep the conditions constant. In the longer term, the mobilisation of fat stores, muscle tone and strength, and bone density are affected by regular exercise, leading to greater fitness.

6.1 The effects of exercise

(a) What physiological effects might someone notice when they start to exercise? **[4]**

(b) Which main body systems are involved in these changes? **[2]**

(c) The table below shows the target heart rate which is ideal for someone to achieve while exercising.
Calculate the target heart rate for someone of 50. **[1]**

	Heart rate in beats per minute	
Age (years)	**Average maximum heart rate (100%)**	**Target heart rate 50–75%**
20	200	100–150
30	190	95–142
40	180	90–135
50	170	
60	160	80–120
70	150	75–113

(d) Why is exercising at a target heart rate desirable? **[2]**

(e) Suggest why the target heart rate is different for different age groups. **[2]**

6.2 Figure 6.1 is a graph showing the oxygen consumption during a period of exercise and rest.

(a) About how long did the period of exercise last? **[1]**

(b) Why does oxygen consumption increase during exercise? **[2]**

(c) If the period of exercise continues long enough, the supply of oxygen may not keep up with demand. What effect does this cause in the muscles? **[2]**

(d) Explain what is meant by an oxygen debt. **[2]**

(e) Why do athletes experience muscle pain when they are exercising particular muscle groups repeatedly? **[2]**

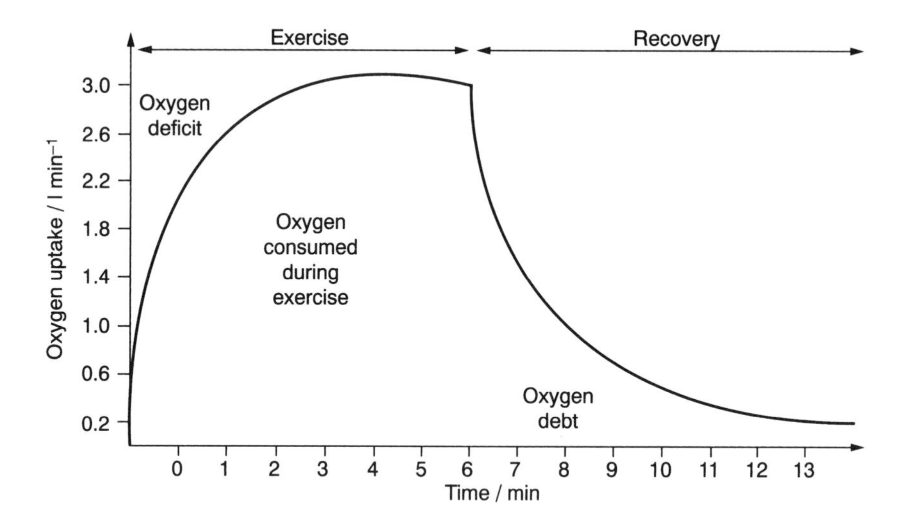

Figure 6.1

6.3 **(a)** What is meant by stroke volume? **[1]**

(b) Write in words the formula for calculating stroke volume. **[1]**

(c) Calculate cardiac output for a person with a stroke volume of 0.105 dm^3, and a heart rate of 66 beats min^{-1}. Is this person likely to be resting or exercising? **[3]**

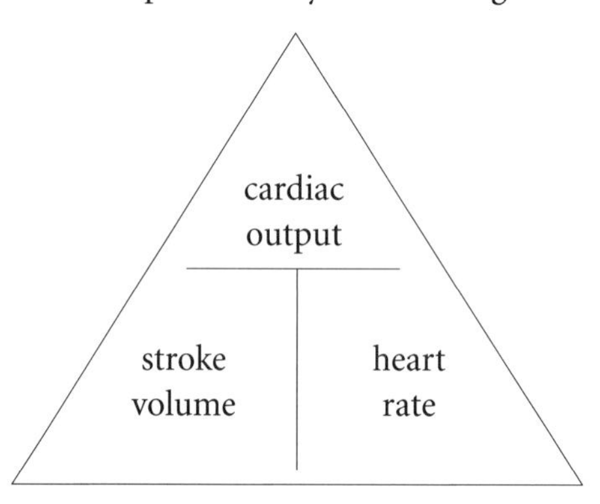

Figure 6.2 Formulas for calculating different aspects of heart function

6.4 Control of breathing rate

Breathing rate varies according to the level of activity the body is involved in. Figure 6.3 represents the system which brings about the control of breathing rate. This system involves sensory elements, a processing component and effectors.

(a) Name two places in the breathing control system where the relevant chemoreceptors are situated. **[2]**

(b) What chemical change stimulates these chemoreceptors? **[1]**

(c) Which part of the breathing centre in the brain receives impulses when these chemoreceptors are stimulated? **[1]**

(d) What type of receptors are situated in the lungs? **[1]**

(e) Which part of the breathing centre are the receptors named in **(d)** connected to? **[1]**

(f) The role of the expiratory centre is to inhibit the inspiratory centre. Why is this necessary? **[1]**

(g) What abnormal conditions might influence this autonomic system? **[2]**

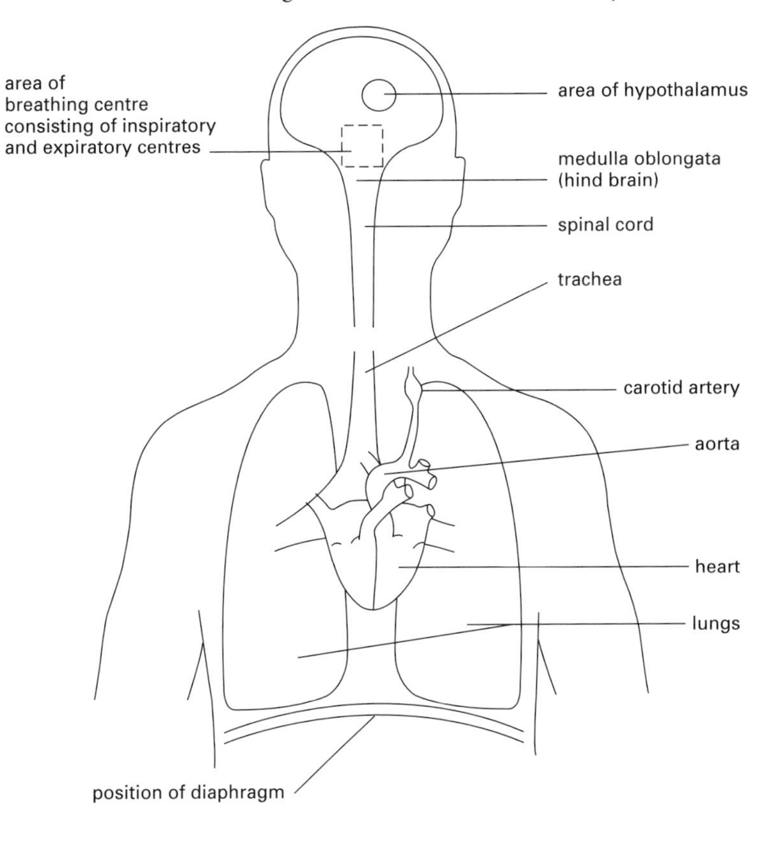

Figure 6.3

6.5 Control of heart rate

(a) Write an account of how the nervous system is involved in controlling heart rate during exercise. Include information about the receptors involved, the stimuli which influence them, and the responses which occur. **[10]**

(b) What part is played by chemical coordination in this control system? **[3]**

6.6 Exercise and fitness

Read this passage about the distribution of blood flow and then answer the questions.

Minimal alterations to blood flow distribution are required during low levels of exercise, although some areas do receive increased circulation. Severe exercise requires massive increase in peripheral flow of blood to the skeletal muscles, along with other physiological changes. These include an increase in heart rate and hence cardiac output (depending on the fitness of the individual), and redistribution of blood flow away from non-essential organs, such as the kidneys.

(a) Suggest which areas of the body might experience slight additional blood flow during low levels of exercise. **[2]**

(b) Why is some blood flow diverted away from non-essential organs? **[1]**

(c) Suggest another body system which might be considered non-essential in this context. **[1]**

(d) Explain why the extent of increase of heart rate during vigorous exercise will depend on the fitness of an individual. **[2]**

6.7 Complete this paragraph by filling in the blanks.

Stamina training increases the ability of mitochondria to generate ATP, which happens during a series of chemical reactions known collectively as __________. More glycogen and fat are stored, and as there is increased __________ in muscles, more oxygen is stored. This means that a fit person who works out regularly relies less on __________ respiration than someone who is unfit, and hence they produce less lactate than an unfit person. Since there is a greater ability to release fatty acids from store, fit people use up more __________ during exercise than unfit people. **[4]**

6.8 The strength of a working muscle partly depends on how extended it is compared to its normal resting length, since this reflects the overlap of actin and myosin filaments and the number of cross-bridges formed between them. Figure 6.4 represents the force of contraction (shown as active tension/% of maximum) at various stages of extension.

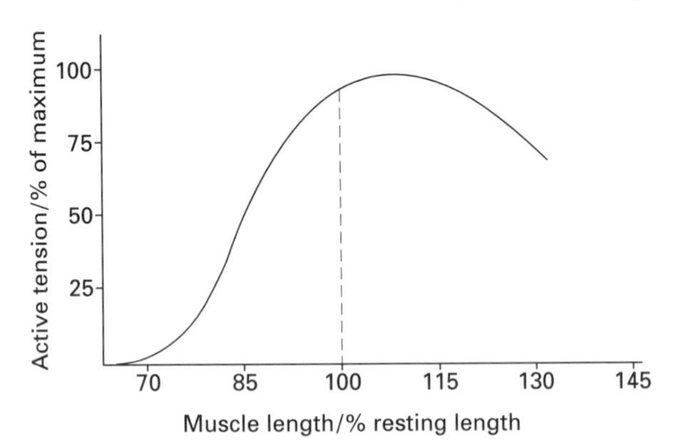

Figure 6.4

(a) What is the muscle extension when it is exerting 50% of its maximum potential tension? **[1]**

(b) At what length compared to its resting length is a muscle able to exert maximum force of contraction? **[1]**

(c) Why does the length of a muscle affect the force of contraction it is capable of? **[2]**

(d) How may the strength of muscles be increased? **[1]**

7 Cell division

During its life a cell goes through the cell cycle. Most cells are in interphase for much of their existence and enter mitosis only in order to divide. Mitotic division produces daughter cells genetically identical to the parent cell. This is important for growth, repair and the replacement of damaged tissues. Cells in the reproductive organs, the testes and ovaries, divide by a different method, meiosis, which results in haploid cells (having half the number of chromosomes as the parent cell). Variation in these cells is increased as a result of the process of crossing over (an exchange of genetic material between homologous chromosomes) that occurs during meiosis.

7.1 Complete the gaps in the following passage, which describes changes in the nucleus prior to cell division.

When a cell is not dividing, the nucleus appears granular when viewed through an optical ______________ . The nuclear material or ______________ consists of __________ bonded to proteins called ______________ . During prophase, strands of chromatin loop and coil, becoming shorter and ______________ . They stain more easily and the ______________ become visible. Each consists of two ______________ held together by a ______________ . **[8]**

7.2 The following statements refer to stages of mitosis and the cell cycle.

(a) Identify the stage described by each statement **and** then give the correct order in which these events occur. **[6]**

- **A** At each pole the chromosomes uncoil. A nuclear membrane forms around each set of chromosomes. The cytoplasm between the daughter cells divides in a process called cytokinesis.
- **B** The centromere of each chromosome divides into two. This separates the chromatids which are pulled to opposite poles of the spindle.
- **C** The chromosomes arrange themselves across the equator of the spindle.
- **D** The cell grows. Within the cell, synthesis of DNA, ATP and proteins takes place. There is an increase in the number of organelles in the cytoplasm.
- **E** Chromosomes, each consisting of a pair of identical chromatids, become visible. The nuclear membrane breaks down and the spindle forms.

(b) In statement **A**, what can you state with some certainty about the 'daughter cells'? **[2]**

(c) In statement **B**, what would happen if the centromere of one chromosome failed to divide successfully? **[1]**

(d) Draw a series of clear, fully labelled diagrams to explain the process of mitosis. Use a diploid cell where $2n = 6$. **[8]**

7.3

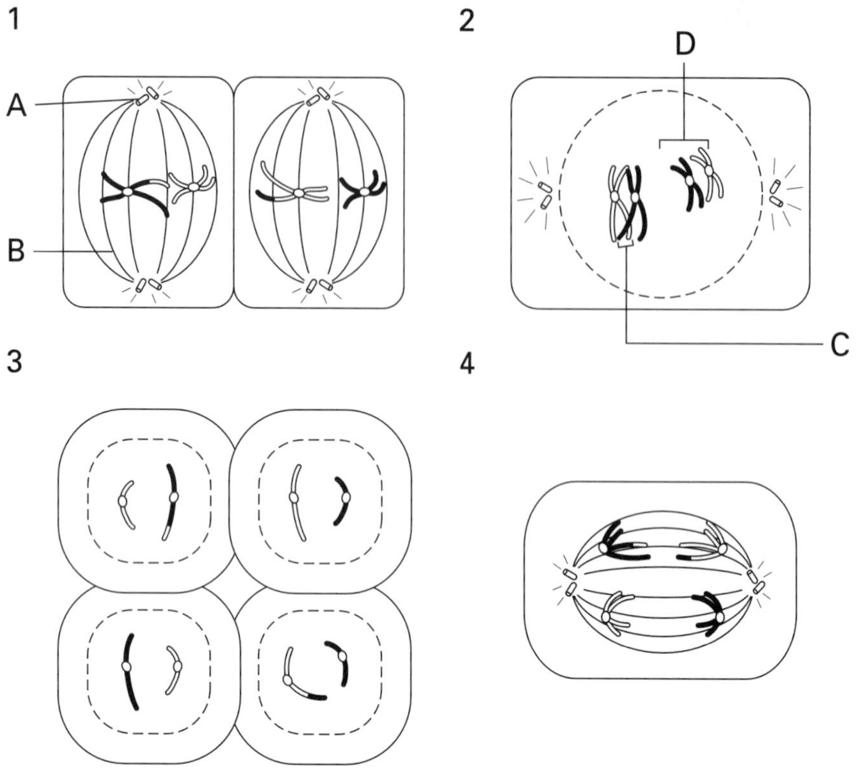

1 2

D

A

B

C

3 4

Figure 7.1 Meiosis

(a) Name precisely the stages of meiosis shown in Figure 7.1. The original cell was diploid and had 4 chromosomes. **[4]**

(b) What is A? **[1]**

(c) What is structure B? What is it made of? **[2]**

(d) What event is occuring at C? Why is it significant? **[3]**

(e) Name D. **[1]**

(f) What is the difference between cell division and nuclear division? **[2]**

7.4

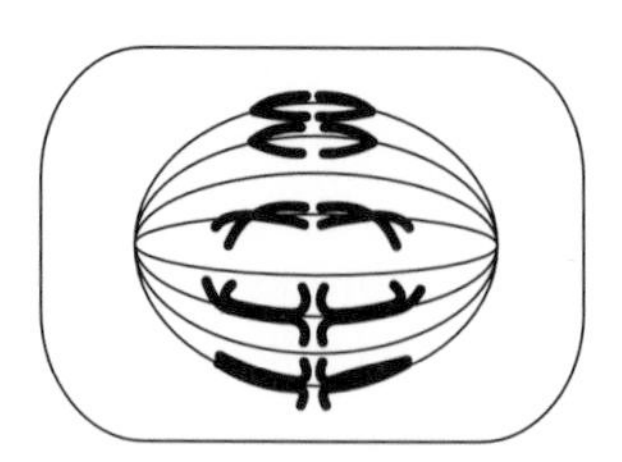

Figure 7.2

This diploid cell is in metaphase 1. It has 10 chromosomes. (2n = 10)

(a) Name the next phase of division for this cell. **[1]**

(b) Draw and label the cell as it will appear towards the end of this next phase. **[4]**

7.5 This table refers to some events which may occur during mitosis and/or the first division of meiosis. Where the event occurs draw a tick in the appropriate box. If it does not occur draw a cross.

Event	Mitosis	Meiosis 1
Chiasmata form and genetic variation may result		
Chromatids separate during anaphase		
Paired homologous chromosomes line up across the spindle equator		
Chromatids become visible during prophase		
The number of chromosomes is halved		
Condensation of chromosomes occurs		

[6]

7.6 Read the following statements and write down for each one whether it refers to mitosis or to meiosis.

(a) The result is two new cells which are identical.

(b) Bivalents line up across the equator of the spindle.

(c) There are two nuclear divisions.

(d) No crossing over resulting in genetic variation.

(e) Four non-identical haploid cells are produced.

(f) It occurs in the Malpighian layer of the epidermis.

(g) It is a process by which genetic variation may occur.

(h) This occurs throughout asexual reproduction.

(i) There may be an exchange of genetic material between homologous chromosomes.

(j) Chromosomes, consisting of two chromatids joined by the centromere, move to opposite poles of the spindle. **[10]**

7.7 **(a)** Explain the terms haploid, diploid, polyploid. **[3]**

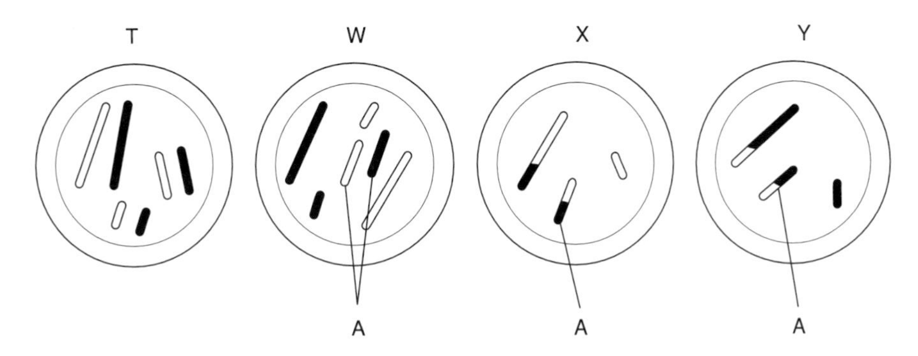

Figure 7.3

Cells W, X and Y have all originated from divisions of cell T.

(b) Which cells in Figure 7.3 are diploid? **[1]**

(c) What is the haploid chromosome number for cells of type T? **[1]**

(d) Which cells, if any, in Figure 7.3 were produced by mitosis? Explain the reason for your answer. **[2]**

(e) Name any of the cells T, W, X or Y that were the products of meiosis? How do you know this? **[2]**

(f) Compare chromosomes A in cells W, X and Y. Explain any differences. **[4]**

(g) Assuming that no crossing over takes place and cell T divides by meiosis, draw four different daughter cells that could be produced. **[4]**

(h) In which two human organs do both mitosis and meiosis occur? **[2]**

(i) The diploid number of chromosomes for humans is 46. How many chromosomes occur normally in the following human cells:

A striped (striated) muscle
B basophil
C squamous epithelial cell
D sperm
E erythrocyte
F cell from a 12 hour morula (embryo)
G primary oocyte
H secondary oocyte **[8]**

7.8 The cell cycle

The cell cycle is a controlled series of events during which a cell grows and divides. Throughout the cycle the quantity of DNA in a cell varies with time as is shown in Figure 7.4.

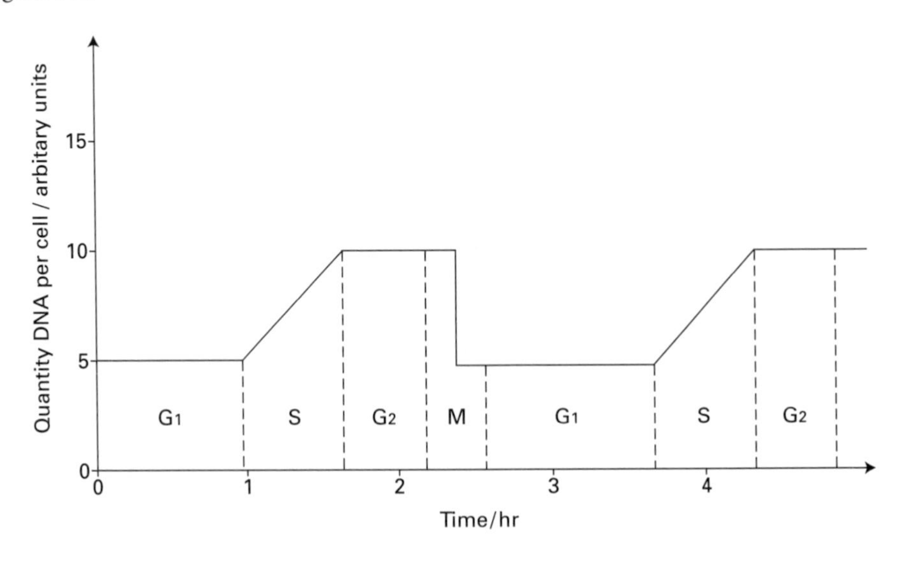

Figure 7.4

(a) Which letter represents the phase of cell division? Explain your answer. **[2]**

(b) During which period is DNA being synthesised? **[1]**

(c) Name one change that can be observed in a cell during phase G1. **[1]**

The cell cycle may also be represented by Figure 7.5

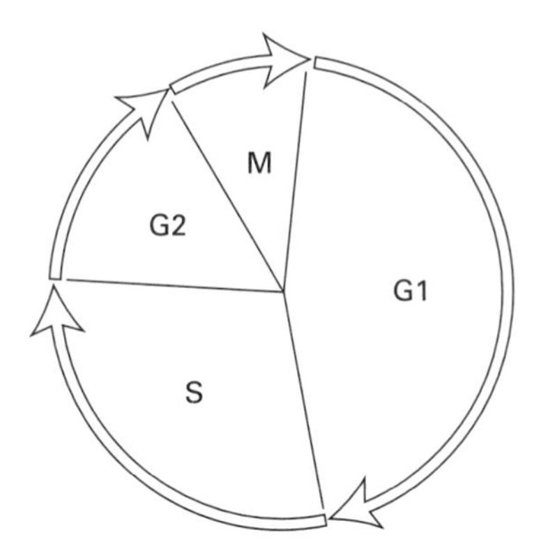

Figure 7.5

During G1 cells receive signals or growth factors which determine whether they remain in G1 or can proceed to S and the rest of the cycle. Human nerve cells, for example, rarely progress beyond phase G1.

If a mutation to cell DNA occurs during G1, the cell should be stopped at phase G2 and not go on to divide by mitosis. This is controlled by a protein p53 which either triggers apoptosis (cell suicide) or reduces the number of times the damaged cell can divide. However, sometimes a mutation occurs in protein p53 and these control mechanisms are disrupted. With the cell cycle out of control the damaged cells proliferate by mitosis; these cells have become malignant.

(d) What terms describe phases G1, S and G2 ? **[1]**

(e) Suggest a second type of human cell that rarely progresses beyond phase G2. **[1]**

(f) Why is it important for DNA to replicate between one mitotic division and the next? **[2]**

(g) Comment on the significance of the observation that p53 protein is absent in 50% of cancer cells. **[2]**

8 DNA and protein synthesis

DNA, 'the double helix', is the nucleic acid found in chromosomes in cell nuclei. It is a polymer made of units called nucleotides. Each nucleotide consists of a molecule of deoxyribose: a phosphate group and one of four nitrogenous bases. DNA is capable of semi-conservative replication. RNA is also a nucleic acid and exists in the cell in three forms. Both are involved in the synthesis of proteins. Triplets of bases along the DNA molecule form the genetic code. This is transcribed by messenger RNA and translated at ribosomes into chains of amino acids called polypeptides. These make up proteins.

8.1 DNA

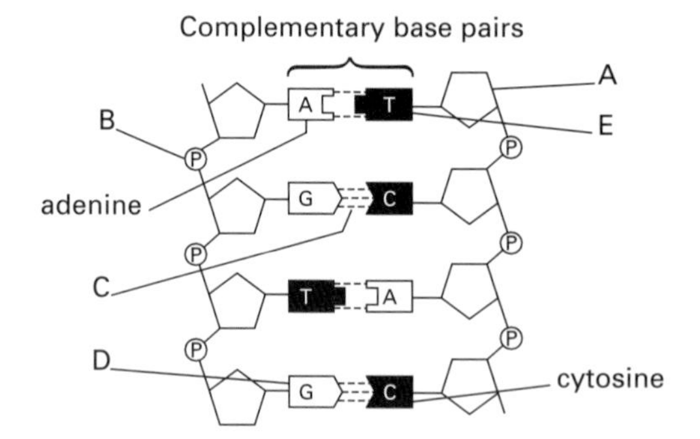

Figure 8.1 Structure of DNA

Copy Figure 8.1

(a) Complete labels A–E. **[5]**

(b) Draw a ring round one nucleotide. **[1]**

(c) Which two nitrogenous bases are purines? **[2]**

(d) Which bases are pyrimidines? **[2]**

In 1951 a scientist called Edwin Chargaff discovered that the number of purine and pyrimidine bases is equal in DNA samples from any living organism. He showed also that the numbers of adenine and thymine bases are equal, as are guanine and cytosine.

(e) What two facts does this show us about the structure of the DNA molecule? **[2]**

8.2 DNA Replication

For some time, scientists have debated how the DNA molecule replicates. One group supported the idea of 'conservative replication'. However, it was eventually proved that DNA replicates by a process called 'semi-conservative replication'.

(a) What is conservative replication? **[1]**

(b) Use the following sentences to write a paragraph explaining how a molecule of DNA replicates semi-conservatively.

- **A** The reactions are catalysed by DNA polymerase enzyme.
- **B** This process is called 'semi-conservative replication.'
- **C** Each new DNA molecule consists of a strand from the original molecule and a new complementary strand.
- **D** These nucleotides join together by condensation reactions to form two new identical DNA strands.
- **E** Nitrogenous bases are exposed along each original strand.
- **F** The DNA molecule unwinds from one end and the strands separate as hydrogen bonds between complementary bases break.
- **G** They act as templates for free nucleotides present in the nuclear sap. **[7]**

8.3 RNA

(a) Copy and complete the following passage:

RNA stands for ________ ________. It is a polymer of ________. Each unit consists of the pentose sugar ________, a phosphate group and a nitrogen base. The bases are ________, ________, ________ and ________. RNA exists in three forms called ________, ________ and ________. **[10]**

(b) How is the pentose sugar in RNA different from the one found in DNA ? **[1]**

(c) Draw and label a molecule of tRNA. **[3]**

8.4 Copy and complete the table to compare DNA and RNA.

	DNA	RNA
Monomer units		
Purine bases		
Pyrimidine bases		
Pentose sugar		
Shape of molecule		
Molecular mass		
Chemical stability		
Where found in cell		
Types of this nucleic acid		

[9]

8.5 The genetic code

(a) Explain what is understood by the 'genetic code'. **[3]**

(b) The following are terms used to describe the genetic code. Write a sentence to define each term.
(i) degenerate, (ii) triplet, (iii) non-overlapping, (iv) universal. **[4]**

(c) Distinguish between the following pairs:
(i) intron, exon
(ii) codon, anticodon **[4]**

(d) This shows the sequence of bases from a strand of DNA which is part of a gene

A	T	G	C	A	C	T	T	T	C	A	C	G	C	T	A	A	G

Write out the sequence of bases on the mRNA that transcribes this part of the gene. **[1]**

(e) The table shows the anticodons of some tRNA molecules and the amino acids that they transport.

Anticodon	Amino acid
GCU	Arginine
CUC	Glutamic acid
AAG	Leucine
UUU	Lysine
AUG	Tyrosine
CAC	Valine

Write the sequence of amino acids that will be found in the polypeptide coded for by the section of DNA shown. **[1]**

8.6 Protein synthesis

(a) Complete the passage that describes protein synthesis.

Protein synthesis begins when a section of DNA unwinds. ____________ bonds between the two strands break. Bases on one exposed strand act as a ______________ and are ______________________ to produce a molecule of ______________. The reaction is catalysed by ________________ enzyme.

mRNA then leaves the ____________ and enters the ________________ where it binds with a ________________. tRNA molecules bring ________ __________ to the site of protein synthesis. Each tRNA ________________ binds with a __________________ codon on the mRNA.

During translation the ____________ moves along the mRNA. As it does so an amino acid ________________ with the ________________ acid at the end of the growing ____________ chain. Energy for protein synthesis is provided by ____________. **[16]**

(b) At the end of translation what happens to:
  (i) the mRNA
  (ii) the tRNA
  (iii) the polypeptide chain **[3]**

(c) What are polyribosomes (polysomes)? **[2]**

(d) Drugs which bind to specific sites on DNA or mRNA molecules can prevent the synthesis of proteins related to some genetic disorders. Triplex drugs (composed of DNA nucleotides) bind with DNA forming a 3-stranded helix. Antisense drugs (made from RNA nucleotides) bind with mRNA.
How do these drugs work with reference to protein synthesis? **[4]**

9 Gene technology

Gene technology involves the manipulation of DNA. In genetic engineering a gene for a useful characteristic can be transferred into the genotype of a different species of organism, often a bacterium, changing its natural phenotype. This procedure has been used to modify bacteria so that they secrete insulin or human growth hormone, which can be produced commercially.

The technology uses enzymes to cut DNA carrying the required gene and to combine it with the DNA of a bacterial plasmid that acts as a vector. The plasmid is inserted into the bacterial host, that is then said to contain 'recombinant DNA'. Plants and animals may also be genetically modified. There is currently debate as to whether this is ethically or practically desirable.

The Polymerase Chain Reaction (PCR) is a technique which allows the very rapid replication of millions of copies of small sequences of DNA. These DNA samples may be collected during forensic investigations, or from historical and archaeological research. Comparison of DNA samples ('DNA or genetic fingerprints') may give proof of relationships between individuals. In a criminal investigation this can be used to prove innocence as well as guilt.

9.1

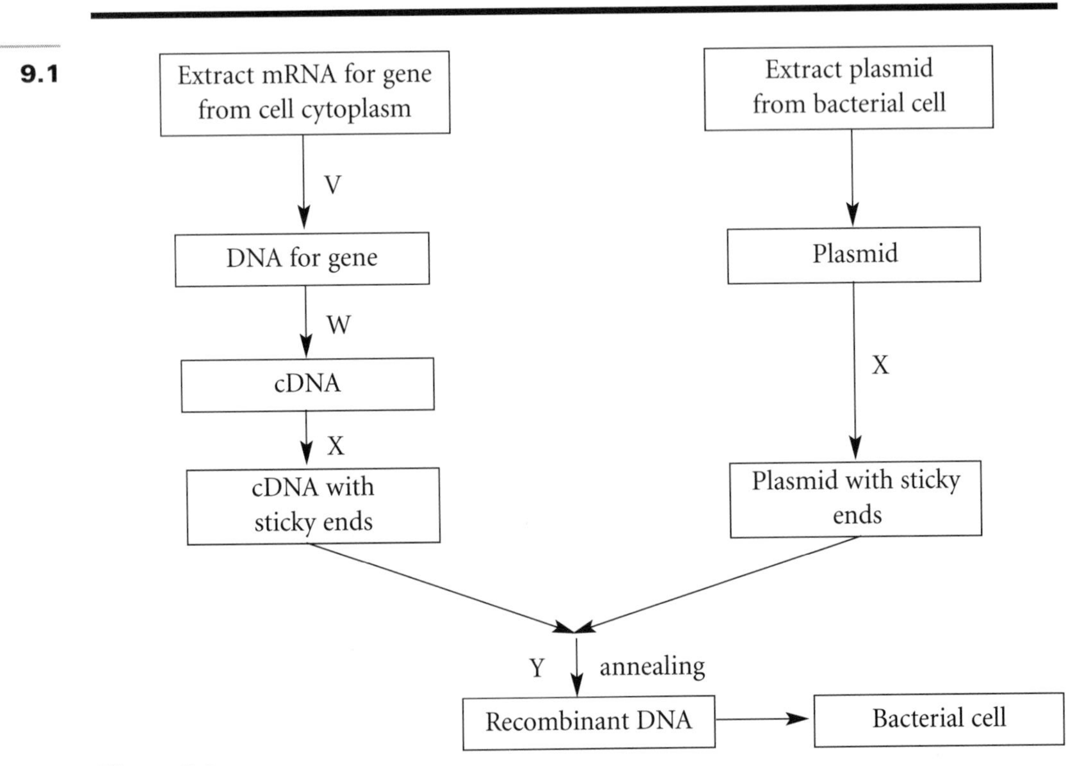

Figure 9.1

(a) What are plasmids? How are plasmids extracted from bacterial cells? **[2]**

(b) Explain the term 'sticky ends'. How do the sticky ends of the plasmid and cDNA form a temporary bond? **[4]**

(c) What is 'annealing'? **[1]**

(d) How is recombinant DNA inserted into bacterial host cells? **[2]**

(e) Complete the following table which relates to the enzymes involved in the process shown in Figure 9.1

Letter	Name of enzyme	Function of enzyme
V		
W		
X		
Y		

[8]

(f) A sample of mRNA extracted from the cytoplasm of a eukaryotic cell contained the following base sequence

U C C A C A G A U U U G

Write the sequence of bases that would be found on the corresponding section of cDNA after treatment with enzyme W. **[2]**

(g) Explain the events that occur once the recombinant DNA has been inserted into the bacterium. **[3]**

(h) Name two products that can be manufactured by recombinant gene technology. **[2]**

9.2 **(a)** What is a marker gene? **[1]**

(b) Plasmids called pBR322 carry genes for resistance to ampicillin and tetracycline. They were used as vectors for a gene that codes for protein X. The insertion of gene X into the plasmid inactivates the gene for tetracycline resistance.

(i) What kind of compounds are ampicillin and tetracycline? **[1]**

(ii) Which of P, Q or R in Figure 9.2 shows the correct position for the insertion of gene X into the plasmid? **[1]**

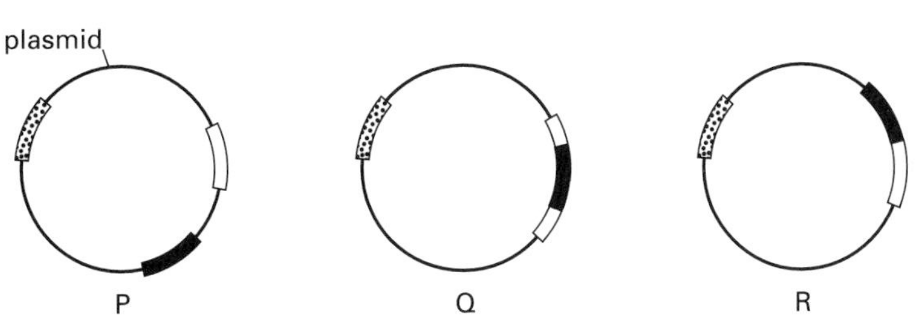

Figure 9.2

(c) Bacteria carrying plasmids from this procedure were plated on to nutrient agar containing ampicillin. A replica plating was made on to nutrient agar containing tetracycline. After incubation for 24 hours at 30°C, the plates showed the following.

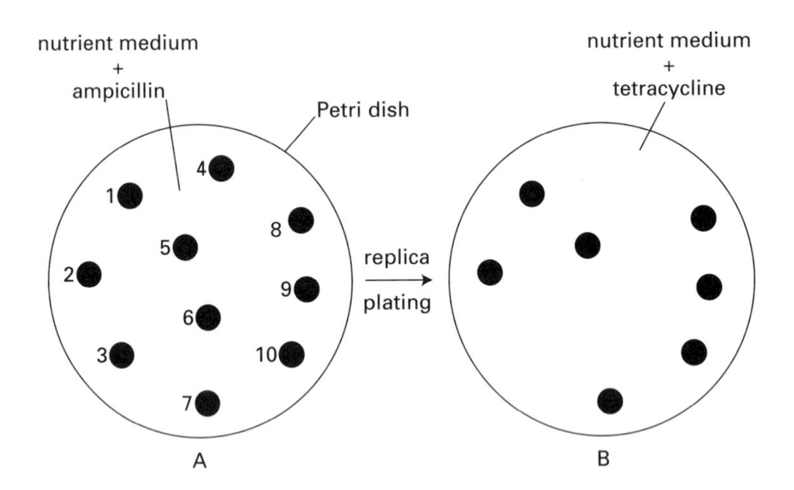

Figure 9.3

(i) What is the origin of each colony? **[1]**

(ii) Which colonies contain bacteria with recombinant DNA suitable for synthesis of protein X? Explain your answer. **[4]**

(iii) How can a pure culture of bacteria carrying gene X be obtained? **[3]**

9.3 **(a)** What is a gene probe and when is it appropriate to use one? **[3]**

(b) The main stages in the use of a gene probe are listed below. Match the following words to the statement to which they most closely relate:

autoradiography hybridisation electrophoresis

A Sample of genetic material is collected.
B DNA is cut into small lengths.
C The shortest lengths of DNA are separated.
D DNA is divided into single strands.
E A chemiluminescent probe is added.
F The probe binds to the complementary base sequence on the DNA.
G The required gene is located. **[3]**

9.4 Genetically modified plants

The soil bacterium *Agrobacterium tumefasciens* invades plants at the junction of root and stem transferring plasmid Ti into the infected cells. This plasmid contains genes for the synthesis of auxins (plant growth substances) and opines (nutrients used by the bacterium). These genes are expressed at a high level leading to increased production of opines and rapid growth of the transformed cells, resulting in plant tumours or 'galls' being formed.

(a) Explain '.....expressed at a high level'. **[1]**

(b) What are 'transformed' cells? **[1]**

(c) What advantage is gained by *Agrobacterium* from this natural process? **[2]**

Geneticists have deleted the genes for auxin and opine synthesis from Ti plasmids, and have instead inserted genes for characteristics potentially beneficial in agriculture.

(d) Suggest three characteristics that might be beneficial if introduced to crop plants. **[3]**

(e) What additional type of gene might be inserted into the Ti plasmid during modification? **[2]**

(f) Why is it necessary for modified Ti plastids to be present in every cell of a transformed plant? **[1]**

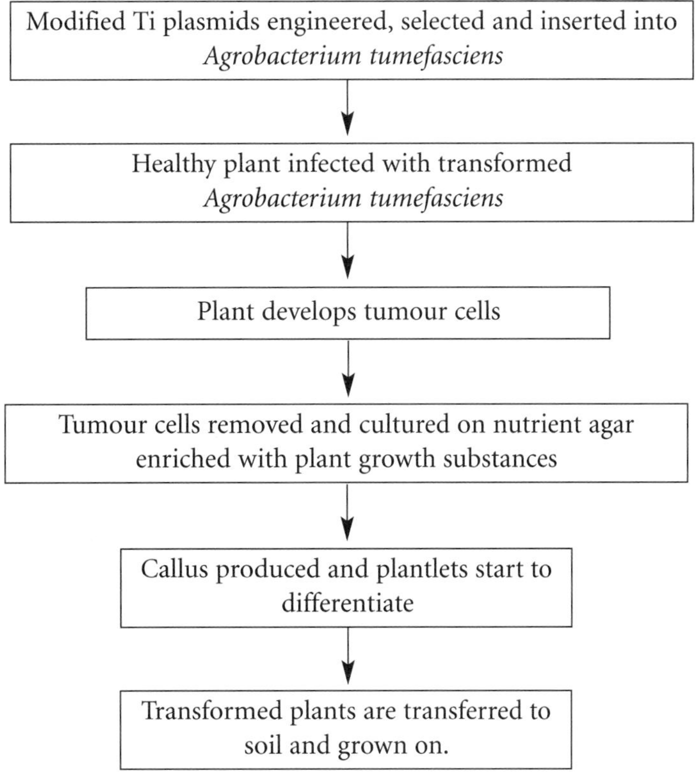

Figure 9.4

(g) List the stages necessary to produce Ti plasmids. **[5]**

(h) What is the vector in this procedure? **[1]**

(i) Name the procedure used to culture plantlets from the callus. **[1]**

(j) What specialised techniques are used during the culture of callus cells? **[1]**

(k) Comment on the fact that *Agrobacterium tumefasciens* does not infect cereal plants. **[2]**

(l) Increasingly, plants are being transformed by firing a tungsten or gold bullet coated with modified DNA (a 'gene gun') into plant cells. How might this be an improvement on the method using *A. tumefasciens?* **[1]**

9.5 Polymerase chain reaction (PCR)

(a) What is the importance of PCR? **[2]**

(b) List four materials necessary to carry out this chain reaction. **[4]**

(c) Copy and complete the following table which summarises PCR:

Stage	1	2	3
Process		annealing	
Approximate time			30 seconds
Temperature °C	90–95		
What happens			2 DNA double helices are formed

[8]

(d) Assuming 100% efficiency, how many copies of the original DNA sequence are obtained after: **(i)** one PCR cycle; **(ii)** ten PCR cycles; **(iii)** twenty PCR cycles? **[3]**

(e) When PCR was first used, the polymerase enzyme was derived from *E. coli* living in the human gut. Fresh enzyme had to be added for each new cycle. Why was this? **[1]**

(f) In the modern PCR procedure thermo-stable enzymes are added at the start of the process. Name a natural source of such enzymes. How else could such enzymes be obtained? **[2]**

(g) Write a paragraph about the applications of PCR. **[4]**

9.6 DNA (genetic fingerprints)

(a) Explain the following terms: **(i)** intron, **(ii)** exon, **(iii)** DNA (genetic) fingerprint. **[4]**

(b) List three ways that DNA fingerprints may be used. **[3]**

(c) The following statements describe briefly the stages involved in the preparation of a DNA (genetic) fingerprint. Put them in the correct order.

A DNA fragments are treated with alkali and separated into single strands.

B Restriction enzymes cut DNA into lengths of between 1000 and 20 000 nucleotides.

C DNA probes hybridise with the DNA strands carrying the required base sequence.

D Electrophoresis is used to separate pieces of DNA by length. **[4]**

Figure 9.5 shows the genetic fingerprints of six individuals.

(d) Identify **(i)** the individuals who are identical twins

(ii) the parents of person B. **[4]**

(e) Explain whether **(i)** D could be the sibling of B

(ii) B could be the sibling of F. **[4]**

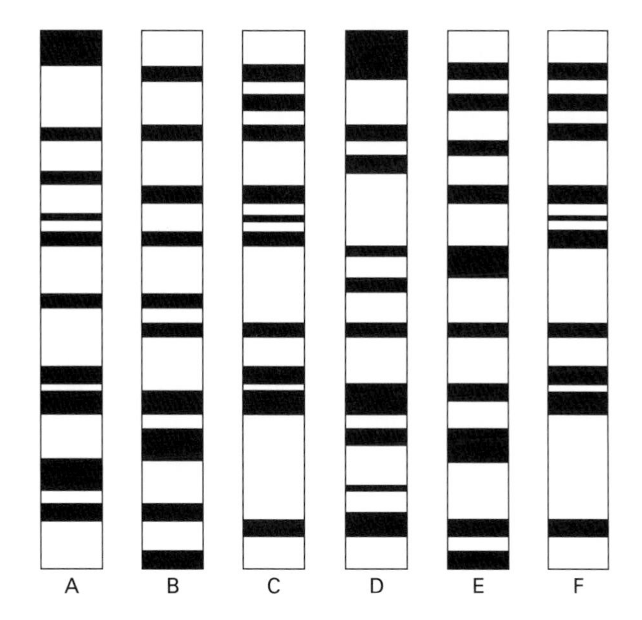

Figure 9.5

9.7 Give an account of the objections put forward by those who do not agree with the growing of genetically modified crops. **[8]**

10 Parasites, pathogens and disease

Parasites rely on other organisms called hosts for vital conditions for life, such as nutrition. The effects of parasites are often very damaging to the host, and may have wide-reaching socio-economic effects for a population. Parasites are adapted to their hosts and have evolved specialised strategies for surviving inside and outside the host. Pathogens are disease-causing organisms which are not necessarily obligate parasites. Many bacteria and viruses are pathogenic.

10.1 Parasitic disease

Copy and complete the table below.

	Tapeworm	Liver fluke	Malaria	Sleeping sickness
Name of parasite				
Name of vector if one occurs				
Name of secondary host if there is one				
Symptoms caused in human host				

[20]

10.2 HIV

Figure 10.1 shows the phases of infection of a human lymphocyte cell by HIV (human immunodeficiency virus).

(a) Describe the components of the HIV particle (virion). [3]

(b) How does an HIV particle enter the lymphocyte? [1]

(c) Why is reverse transcriptase (stage 1) vital to the success of HIV? [2]

(d) Viral DNA integrates with the host cell DNA. What is the name of the process by which mRNA is copied from DNA (stage 2)? [1]

(e) mRNA acts as a code for the production of proteins and more viral RNA within the cytosplasm of the host. What is the name of this process (stage 3)? [1]

(f) What is the effect of release of viral particles on the host cell? [2]

(g) How is HIV able to escape the immune system of the host? [2]

(h) What is the long term effect of HIV on the immune system? [1]

(i) Suggest how the cycle of production of virions might be interrupted. [3]

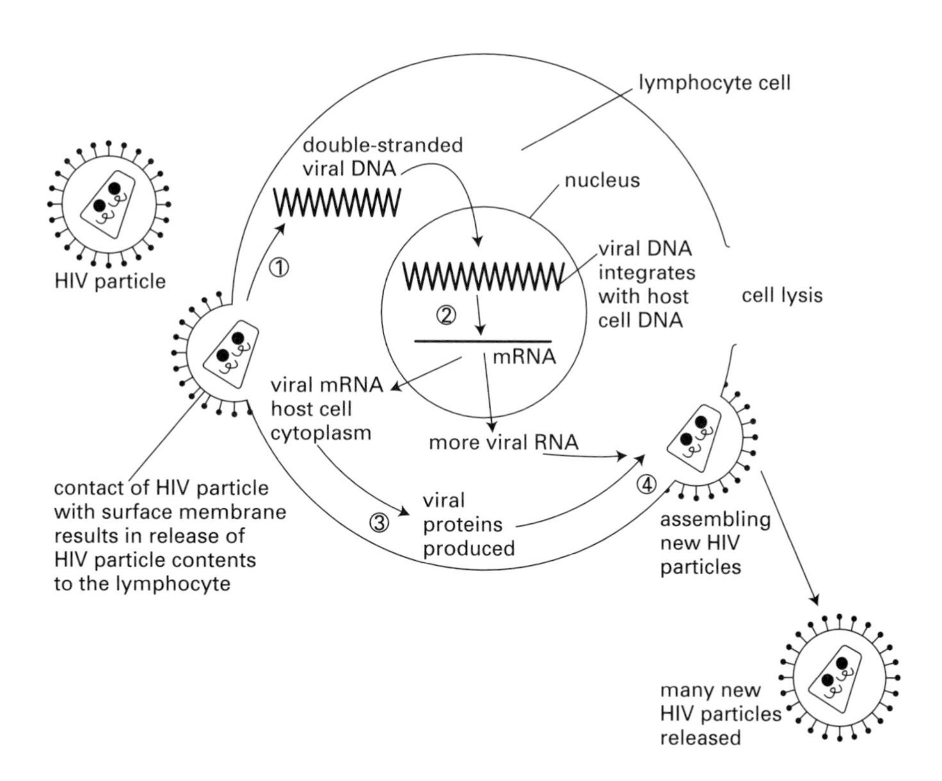

Figure 10.1

10.3 Bacterial population growth

On reaching a certain size, bacteria divide by a type of asexual reproduction. This can happen very rapidly. In the table, the time taken for a bacterium to grow and divide once to form two new cells, is 30 minutes.

	Time (1 unit = 30 min)									
	1	2	3	4	5	6	7	8	9	10
A Number of bacteria										
B Log_{10} number of bacteria										

(a) Copy the table and starting with the first generation of one bacterium, complete line A in the table, assuming one division every 30 minutes. **[1]**

(b) Complete line B by finding $\log_{10}$ for the number of bacteria at each generation. **[1]**

(c) Plot line B on graph paper, using $\log_{10}$ number of bacteria on the y-axis and time on the x-axis. **[5]**

(d) What is the advantage of using $\log_{10}$ of the number of bacteria to display the data? **[1]**

(e) Explain what is meant by exponential, by reference to line A on the graph. **[2]**

10.4 A student investigated the population of bacteria on the skin of people's hands following washing and drying. The same washing method was employed, but two drying methods were used: paper towel and hot air blower. Swabs were used to take samples from skin and bacteria cultured from the swabs.

Bacterial populations on hand skin following washing and drying

Sample	Air-dried skin/$\times 10^8$cm^{-2}	Towel-dried skin/$\times 10^8$cm^{-2}
1	8.91	1.11
2	9.75	0.98
3	6.14	0.42
4	8.72	1.02

(a) What main trend is demonstrated by these results? **[2]**

(b) Suggest a factor which is unlikely to have caused the difference between the two groups, and one that is likely to have a significant effect. **[2]**

(c) Why might the factor you have described in **(b)** make a significant difference? **[2]**

(d) Suggest two other conditions which would need to be controlled to make this a fair test. **[2]**

10.5 The spread of disease

Figure 10.2 shows a simple representation of the Malaria life cycle.

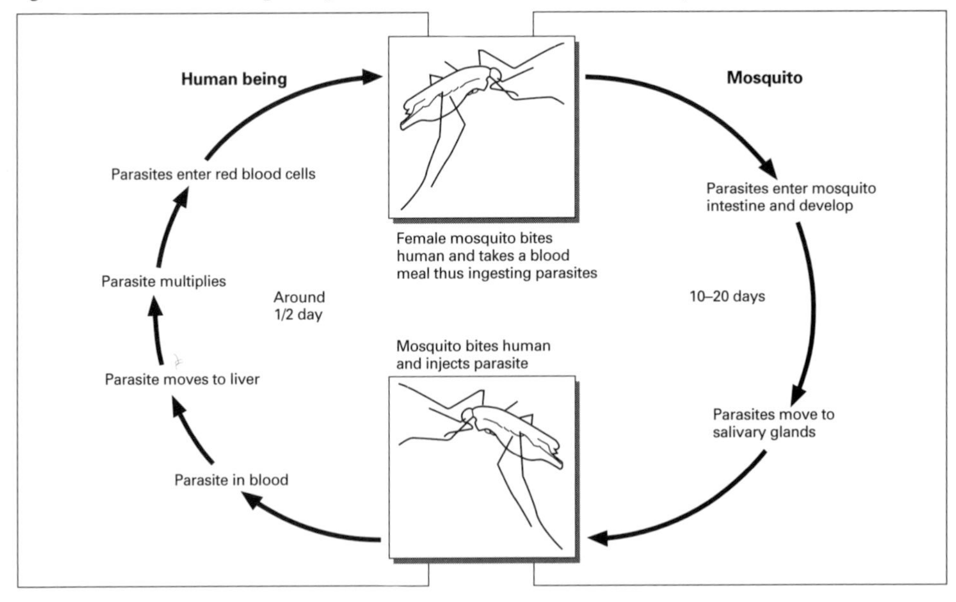

Figure 10.2

Using this diagram for information and your knowledge of the disease, suggest varied strategies for preventing deaths from malaria. **[5]**

10.6 This is an abstract written by J K Baird of the US Naval Medical Research Unit No 2, Jakarta, Indonesia, published in April 2000. Read the article and answer the questions below.

... In 1955 the World Health Organisation (WHO) adopted the controversial Global Eradication Campaign emphasising DDT ... spraying in homes. The incidence of malaria fell sharply where the programme was implemented, but the strategy was not applied (throughout) Africa. This, along with the failure to achieve eradication in larger tropical regions, contributed to disillusionment with the policy. The WHO ... abandoned the eradication strategy in 1969. A resurgence of malaria began at about that time and today reaches into areas where eradication or control had been achieved. ... In 1993 the WHO adopted a Global Malaria Control Strategy that placed priority in control of disease rather than infection ... Emphasising diagnosis and treatment in the primary healthcare setting ... and de-emphasising the spraying of residual insecticides ... Resurgent malaria accompanying declining vector control activities in Asia and the Americas suggests that (this) may be premature. ... The inadequacy of vector control as the primary instrument of malaria control in ... Africa does not preclude its utility in Asia and the Americas.

(a) Which part of the malaria life cycle is interrupted by DDT spraying? **[1]**

(b) What was the initial outcome of the Global Eradication Campaign in 1955? **[1]**

(c) Why was this campaign subsequently abandoned? **[2]**

(d) How did the Global Malaria Control Strategy adopted by the WHO in 1993 differ from the 1955 campaign? **[3]**

(e) What is a **residual insecticide** and what are some of the concerns over its use? **[2]**

(f) Why might discontinuing the use of vector control activities be premature? **[2]**

10.7 *Since vaccination for polio (poliomyelitis) began in 1955 the disease has been all but eradicated (except in a few regions of Asia and Africa). The WHO has an aim to eradicate polio completely by 2005 after which vaccination could stop, saving $1.5 billion annually.*

The most common form of vaccine is an oral suspension (called OPV), containing a live but weakened form of the virus. The earlier form of vaccine (called Salk), which contains killed polio virus, has to be injected. The oral version is cheaper and easier to administer and the viruses from it can spread to unvaccinated people offering them protection.

OPV viruses differ from polio viruses by only a few gene mutations, and it is possible that they could mutate again to more dangerous forms. There is still debate about whether vaccination should continue for some years after eradication to reduce the danger of reinfection from mutated forms of virus.

Read the passage and answer the questions.

(a) How has the OPV vaccine helped towards nearly eradicating polio? **[3]**

(b) What are two main advantages of total eradication? **[2]**

(c) Describe two differences between the OPV vaccine and Salk vaccine. **[2]**

(d) Suggest a reason why some scientists think it is important to continue to vaccinate after eradication. **[1]**

(e) How would continuing to vaccinate using OPV affect the global pool of polio-type viruses? **[1]**

10.8 **(a)** Which bacterium causes tuberculosis (TB)? **[1]**

(b) Which type of medicine mostly accounts for the success in controlling TB? **[1]**

(c) The rate of infection with TB is increasing in developing and developed countries. Explain why it is not possible to treat these new cases of TB with the same medicines first used to control the disease. **[2]**

10.9 Write an account about the ways in which pathogens and parasites enter the human body. **[5]**

10.10 The World Health Organisation has recognised the need for measuring haemoglobin (Hb) in small rural health centres, because anaemia is an indicator for many health conditions such as parasite infestation. The following data are taken from a study of Peruvian Indians living in isolated rural areas.

Age group	Intestinal parasites	Hb (g/l) Mean	Hb (g/l) Range of central 50%
<1–5 years	yes	96.6	87–107
	no	97.1	90–105
6–14 years	yes	103.6	94–113
	no	109.0	102–116
15 years and over	yes (males)	119.5	108–139
	no (males)	125.1	118–135
	yes (females)	107.3	98–120
	no (females)	108.9	98–120

(a) What is the trend shown by these data for the level of Hb as people grow older? **[2]**

(b) From the means shown here, suggest whether the presence of intestinal parasites influences the level of Hb in blood, and explain your answer. **[2]**

(c) Why is the data for central range useful in this type of study? **[1]**

(d) Suggest why the researchers considered the data for males and females separately in the last age category. **[2]**

(e) Explain how mean Hb levels might differ between this population and a group surveyed in Europe, and why. **[2]**

10.11 The incidence of meningococcal disease (meningitis) amongst teenagers and young adults in developed countries increased during the 1990s. Different strains of meningococci occur, some of which may be potentially life-threatening. A study carried out at Nottingham university provided information about the carriage and acquisition rate of *Neisseria meningitis* (*N. meningitidis*) in first year students.

Date	% carriage rate
30 September	6.9
1 October	11.2
2 October	19.0
3 October	23.1

Carriage rate of *N. meningitidis* during the first week of term

Factor	Odds ratio (95% confidence interval)
Passive smoking	1.21 (0.9–1.7)
Smoker	1.6 (1.0–2.6)
1–4 weekly visits to bar	1.74 (1.1–2.8)
> or = 5 weekly visits to bar	2.71 (1.5–4.8)
Male only hall of residence	0.71 (0.5–1.0)
Female only hall of residence	0.52 (0.3–0.9)
Self-catered hall of residence	0.73 (0.5–1.2)

Some risk factors for acquisition of *N. meningitidis*

(a) What is meant by these terms:
  (i) acquisition? **[1]**
  (ii) carriage? **[1]**

(b) Suggest reasons why the patterns in carriage of microflora might be expected to change when new students arrive at a university. **[5]**

(c) What is the main trend in data for the first week of term? **[2]**

(d) The study concludes that social interaction is important in acquisition rate of *N. meningitidis*. Which data in the second table indicate this? **[2]**

(e) What other risk factors can be identified from these data? **[3]**

(f) Why are statistical tests useful in analysing data? **[1]**

(g) What is meant by 95% confidence interval? **[1]**

10.12 Some infectious diseases are notifiable by law under the Public Health (Infectious Diseases) Regulations 1988. The table below shows some of the data for England and Wales.

Disease	**1990**	**1991**	**1992**	**1993**	**1994**	**1995**	**1996**	**1997**	**1998**	**1999**
Poliomyelitis	1	4	3	1	0	1	1	2	2	0
Cholera	19	22	25	23	30	32	32	33	48	29
Food poisoning	52145	52543	63347	68587	81833	82041	83233	93901	93932	86316
Measles	13302	9680	10268	9612	16375	7447	5614	3962	3728	2438
Rabies	0	0	0	0	0	0	0	0	0	0
Smallpox	0	0	0	0	0	0	0	0	0	0
Tetanus	9	8	6	8	3	6	7	7	7	3
Viral hepatitis A and B	9005	8860	8993	5557	3722	3296	2437	3186	3183	3424
Whooping cough	15286	5201	2309	4091	3964	1869	2387	2989	1577	1139

Notification of infectious diseases 1990–1999

(a) Suggest reasons why some diseases are notifiable by law while others are not, and give an example of the latter. **[4]**

(b) The term food poisoning is a general term covering illness caused by several different causative agents. What are the common ways in which food poisoning is transmitted? **[3]**

(c) Suggest a reason related to lifestyle which might account for the rise in food poisoning in the 1990s. **[2]**

(d) A vaccine for poliomyelitis is routinely offered to all parents for newly born and young children. Suggest reasons why some cases still occur. **[2]**

(e) The number of cases of measles has dropped significantly since 1994. Suggest how health education might have contributed to this trend. **[2]**

(f) No cases of rabies or smallpox are recorded in this table, but the reason is different in each case. What are the reasons why rabies and smallpox have not been notified in England and Wales during this period? **[2]**

(g) Routine immunisation for cholera is not offered in the UK. Suggest a reason why not. **[1]**

(h) Under what circumstances is cholera immunisation recommended to people? **[1]**

11 Heart disease

Cardiovascular disease is the world's leading cause of death and is still increasing in many countries. The main health problems are hypertension (high blood pressure) and stroke, coronary heart disease (leading to heart attacks) and circulatory disease. Many lifestyle factors influence heart health and disease, such as body mass, exercise, diet and smoking, as well as familial genetics. Both medicines and surgery may be used to treat cardiovascular problems.

11.1 **(a)** Explain what is meant by:

(i) hypertension **[1]**

(ii) stroke **[2]**

(iii) coronary heart disease **[2]**

(b) In many developed and developing countries the rate of death from coronary heart disease is around 50%. Suggest reasons why this rate increases as a country becomes more developed. **[3]**

(c) Which gender is more affected by coronary heart disease in developed countries? **[1]**

11.2 Copy and complete the following table of factors which contribute to hypertension.

Factor	Reason why it affects the cardiovascular system
High salt intake	____________ (1)
Smoking	____________ (2)
____________ (3)	inherited characteristics which predispose to hypertension
Obesity	____________ (4)

[4]

11.3 Define these terms:

(a) angina **[1]**

(b) atheroma **[1]**

(c) thrombosis (thrombus) **[1]**

(d) stroke **[1]**

(e) angioplasty **[1]**

(f) aneurism **[1]**

11.4 What is the effect of hypertension on

(a) the walls of larger blood vessels? **[1]**

(b) heart function? **[2]**

(c) the kidney? **[1]**

(d) smaller vessels such as capillaries? **[1]**

11.5 The table below shows the results of a study in Chicago of 1822 men aged 40 to 55. Cause of death and fish consumption (g/day) were recorded. All the men were free of cardiovascular disease at baseline. There were 430 deaths from coronary artery disease during the study period, of which 293 were from heart attacks.

Fish consumption g/day	Death rate from heart attack (%)
0	78.6
1–17	69.2
18–34	57.9
>35	45.3

(a) Apart from a heart attack, why else might coronary artery disease cause death? **[2]**

(b) What main trend appears to be shown by the data in the table? **[1]**

(c) What other factors might affect the coronary health of individuals in the study group? **[5]**

(d) What scientific evidence is there that fish oils increase coronary health? **[2]**

11.6 **(a)** Write a list of self-help hints which might help decrease the incidence of hypertension in a population. **[5]**

(b) Suggest reasons why a government would be interested in reducing the incidence of coronary heart disease? **[4]**

(c) Which heart problems are most likely to affect children? Explain why. **[3]**

11.7 **Treating heart disease**

Match each of the following treatments: **beta-blocker, diuretic, calcium channel blocker** to the explanation of how it works from the list below.

A Affects arteriole muscle, relaxing and widening vessels.

B Slows heart rate and reduces heart output.

C Increases salt excretion, so lowering water potential in tissues **[3]**

11.8 Recently, around 2000 heart transplants have been carried out in the US every year, although more than 20 000 Americans might benefit from this procedure. Of the transplants carried out, around 75% are in men and around 52% are in patients aged 50–64. The one-year survival rate is 83%, two-year survival rate is 78%, the three-year survival rate is 74% and the four-year survival rate is 71%.

(a) Why are fewer transplants carried out than the number of people who might benefit from this technique? **[2]**

(b) Suggest a reason why more males receive a heart transplant than females. **[1]**

(c) What trend is shown by the data for survival rates? **[2]**

(d) What is the main reason for failure of heart transplants? **[1]**

11.9 Monitoring heart function

Figure 11.1 represents the electrical activity of the heart as it beats.

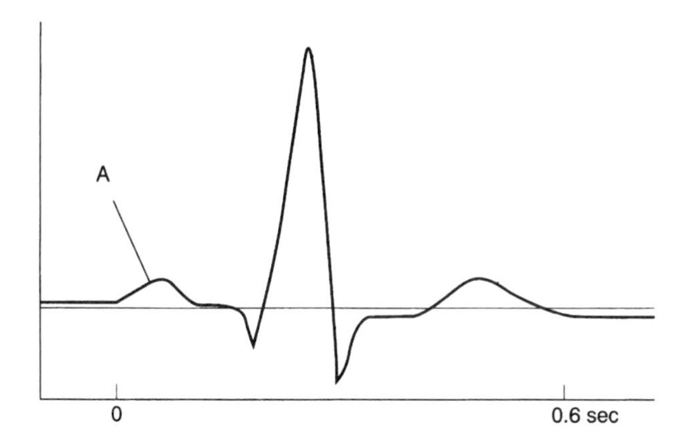

Figure 11.1

(a) What is this type of diagrammatic record called? **[1]**

(b) What electrical activity within the heart corresponds to the point marked A on the diagram? **[2]**

(c) Which of the points on the diagram represents the depolarisation of the major part of the heart muscle? **[1]**

(d) What is meant by repolarisation, and when does it happen? **[3]**

11.10 Figure 11.2 shows the pressure changes that occur in the right side of the heart and associated artery.

(a) What is the highest pressure exerted by each heart chamber during the cycle? **[2]**

(b) Why is there a difference in the pressure exerted by these two chambers? **[2]**

(c) What is the role of pulmonary artery? **[1]**

(d) Explain why the blood pressure is greatest in the right ventricle while it is lowest in the right atrium. **[2]**

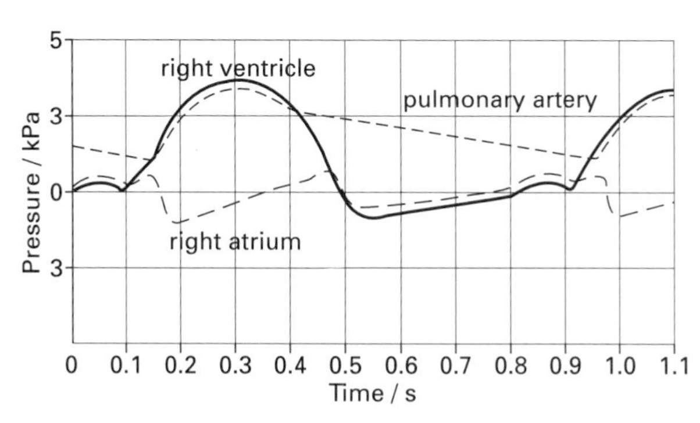

Figure 11.2

12 Immunity, diagnosis and disease

Immunity is a system of body defence which offers protection against disease caused by infection. The immune system can detect the difference between your own (self) cells and non-self cells by recognising unique protein markers which exist at cell surfaces. The immune response is a set of reactions mostly brought about by lymphocytes to counteract infection. The main mechanisms for doing this are producing antibodies, engulfing non-self particles and neutralising toxins.

The skin plays a part in body defence as it is a physical barrier to many substances and to microbes. The blood system helps to maintain this barrier, as clotting and scab formation occur during wound healing.

There are many techniques for diagnosing disease, including monitoring the body's vital signs (such as blood pressure), measuring other specific parameters (such as level of plasma proteins or ions in blood), and using antibodies in test kits (ELISA) and DNA probes.

12.1 **Immunity**

Define the following terms:

(a) antigen **[1]**
(b) agglutination **[1]**
(c) lymphocyte **[1]**
(d) antibiotic **[1]**
(e) bacteriostatic **[1]**
(f) epidemiology **[1]**
(g) immunoglobulin **[1]**

12.2 Copy and complete the following passage by writing appropriate words in the blank spaces.

Phagocytosis occurs at the site of an injury when ____________ such as bacteria enter the tissues. The ____________ recognise the antigens as foreign to the body, and respond by ____________ blood capillaries to pass between tissue cells. Their ____________ flows to alter the shape of the cell, allowing it to ____________ an antigen by surrounding it. The antigen is held within a ____________ vesicle into which digestive enzymes are secreted. These enzymes break down the ____________, destroying it and preventing further infection. **[7]**

12.3 The skin provides a barrier between tissues and body surroundings.

(a) Name and briefly describe the main layers of skin starting at the surface and moving towards deeper layers. **[5]**

(b) Which tissue provides elasticity to skin, helping to prevent splitting, and why is this important for health? **[2]**

(c) How does the skin surface act as a barrier to infection? **[3]**

(d) Sweat glands form an important part of the body's thermoregulation mechanism. This is because energy is needed for sweat to evaporate. Calculate the percentage of energy used by an athlete who has a daily energy intake of 48 000 kJ, and loses 3.9 litres of sweat. (The latent heat of evaporation of sweat is 2.47 kJ cm^{-3}). **[3]**

12.4 Copy and complete the following table about the roles of lymphocytes, putting a tick where a statement is true and a cross where a statement is false.

Feature	T-lymphocyte	B-lymphocyte
Produced from stem cells which arise in bone marrow		
Help agglutinate antigens		
Engulf antigenic particles by phagocytosis		
Produce antibodies		
Bind to antigens, destroying them		
Form plasma clone cells		

[12]

12.5 **(a)** Suggest a reason why plasma clone cells produced by B-lymphocytes are only active for a few days, after which they die. **[1]**

(b) Immunoglobulins are protein molecules in which part of the structure is variable and produced specifically to match an antigen. Briefly describe why this specificity is key to how antibodies work. **[3]**

(c) What is meant by immune memory and why is it important to health? **[2]**

(d) Figure 12.1 shows antibody levels in blood serum after exposure to a particular antigen.

IgM and IgG are two different classes of immunoglobulins.

(i) Which immunoglobulin is important in the body's first response to an antigen? **[1]**

(ii) Describe the pattern of immunoglobulin production when a second exposure to a particular antigen occurs. **[2]**

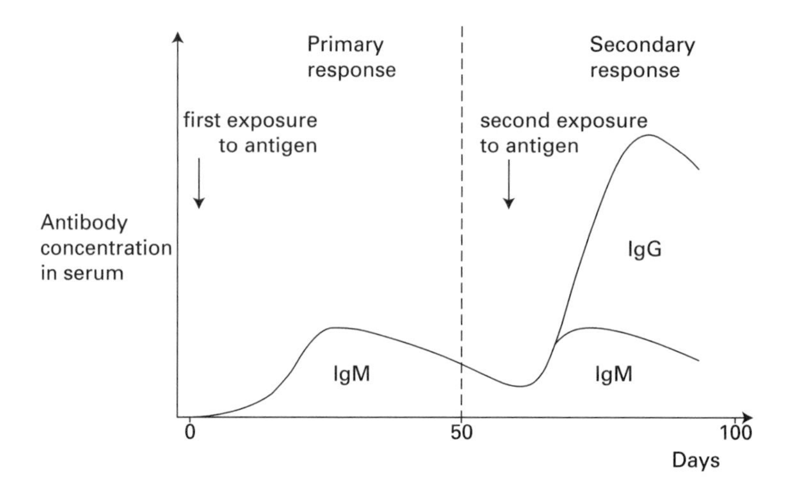

Figure 12.1

12.6 Give one example of each of the following types of immunity:

(a) natural active immunity **[1]**

(b) natural passive immunity **[1]**

(c) acquired active immunity **[1]**

(d) acquired passive immunity **[1]**

12.7 Write a short account of the influence of the development of antibiotics and immunisation on the patterns of health and disease of a population. **[10]**

12.8 Diagnosing disease

Enzyme levels in plasma can be used in diagnosis of disease or organ injury. For example, enzymes such as amylase and creatine kinase occur in cell cytoplasm or mitochondria, and small amounts are readily discharged into plasma when minor damage or alteration to cell membranes occurs. Other enzymes, such as alkaline phosphatase, are bound to membranes within cells.

When a single event involving damage or disease occurs, the level of an enzyme in plasma first increases and then declines as it is removed from circulation. If the condition is prolonged or progressive, the removal of enzyme from plasma may not be as fast as its leakage into plasma.

(a) Read the information above and name an enzyme which will not normally appear in cell fluid. **[1]**

(b) How might an enzyme be removed from plasma by the body? **[1]**

(c) In what circumstances might the level of an enzyme associated with damage or disease remain constant over a period of time? **[1]**

The table below shows the half life of some enzymes in dogs and cats.

	Half life of enzymes (t½) (hours)	
Enzyme	**Dog**	**Cat**
Alanine aminotransferase	48	48
Amylase	12	12
Aspartate aminotransferase	12	40
Lactate dehydrogenase	18	30
Lipase	18	18

(i) If the concentration of alanine aminotransferase in a feline plasma sample is 100 units, how long will it be before the concentration drops to 50 units (assuming no further production)? **[1]**

(ii) The concentration of lactate dehydrogenase in a sample of feline plasma and a sample of canine plasma is 40 units, and no more production is occurring. How much longer will it take the feline lactate dehydrogenase to reach 10 units than the other sample? **[2]**

(e) The table shows components in plasma as indicators of urinary malfunction.

Measured plasma component	Chronic renal failure	Acute renal failure	Urinary tract obstruction
Albumen	↓	⬇	
Amylase	↑	↑	↑
Creatinine	⬆	⬆	⬆
Inorganic phosphate	↑	↑	↑
Total protein	↓		
Urea	⬆	⬆	⬆

(i) Albumen is a protein. Explain why it is not normally found in urine. **[2]**

(ii) Why might the level of albumen in plasma drop if the kidneys are malfunctioning? **[2]**

(iii) What is the source of urea in urine? **[1]**

(iv) Why might the level of urea in plasma rise if the kidney is damaged? **[2]**

(v) How might urinary tract obstruction be distinguished from renal failure using these test results? **[2]**

(vi) Why would these tests be used alongside other diagnostic methods? **[1]**

12.9 ELISA stands for enzyme-linked immunosorbent assay, which involves both antibodies and enzymes. Briefly describe how this technique works. **[5]**

13 Inheritance

Every individual has a unique phenotype, which is the result of interaction between the environment and its genotype. Geneticists study the patterns of inheritance of characteristics passed from one generation to the next. A gene occupies an exact position, its locus, on a chromosome. It exists in different forms, alleles, which are expressed in the phenotype as the dominant or recessive characteristic. The recessive characteristic shows only if the individual is homozygous for the recessive allele.

Monohybrid inheritance is the inheritance of a single characteristic. If two individuals heterozygous for the characteristic are crossed, the phenotypes of their offspring show the ratio 3:1, dominant:recessive. Sometimes two alleles of the same gene both contribute to the phenotype of the offspring (neither allele is totally dominant over the other) and a third phenotype is produced. These alleles are co-dominant.

Dihybrid inheritance involves two genes carried on different chromosomes. Crossing two heterozygotes produces offspring with four possible phenotypes in the ratio 9:3:3:1. Gregor Mendel first discovered these patterns. Linked genes, that is two genes occurring on the same chromosome, tend to be inherited together. Their offspring do not show the 9:3:3:1 ratio.

Genes located on the X sex chromosome are described as 'sex-linked'. These often result in the recessive disorders; for example, colour blindness is seen more frequently in males. Some genetic diseases are due to a single dominant allele, e.g. Huntington's Disease, but others, e.g. Cystic Fibrosis, are due to individuals being homozygous for the recessive allele.

13.1 In brief sentences explain the difference between the following terms:

(a) chromosome, gene, allele. **[4]**

(b) dominant gene; co-dominant gene. **[2]**

(c) homozygote, heterozygote. **[2]**

(d) monohybrid inheritance, dihybrid inheritance. **[2]**

In tomato plants a single gene controls the shape of the fruit. A tomato plant homozygous for round fruit was crossed with a plant homozygous for oval fruit. All the F_1 plants had round fruit. When the F_1 plants were crossed, the plants in the F_2 generation consisted of 75% plants producing round fruit and 25% producing oval fruit.

(e) Which characteristic for fruit shape is dominant? How do you know this? **[2]**

(f) Using R for the dominant allele and r for the recessive allele for fruit shape, give the genotypes of:

(i) the parent with round fruit

(ii) the parent with oval fruit

(iii) the F_1 plants **[3]**

(g) Draw a Punnett square to show the genotypes and phenotypes obtained when the F_1 plants were crossed. **[4]**

(h) What percentage of the F_2 generation were heterozygous plants? **[1]**

13.2 A healthy couple had five children. Two suffered from phenylketonuria (PKU), a metabolic disorder caused by a recessive allele; the other three children were healthy.

(a) Using P for the dominant allele and p for the recessive allele, draw a genetic diagram to show how healthy parents can have a child with PKU. **[3]**

(b) What is the chance of any child of this couple having PKU? **[1]**

(c) What are the possible genotypes of healthy children born to this couple? **[2]**

(d) What is the probability of a healthy child in this family being a carrier for PKU? **[2]**

(e) Explain by means of a diagram, the probability of a PKU carrier and a partner who is homozygous for the normal condition having a child with PKU. **[2]**

(f) Most babies born in the UK are tested for PKU within a few days of birth. How is this done and what is being looked for? **[2]**

(g) Cystic fibrosis (CF) is another condition caused by a recessive allele. Write a short paragraph about the symptoms and problems of cystic fibrosis. **[4]**

(h) What treatments are currently available to sufferers of CF? **[2]**

(i) What treatment may be available in the future for CF? **[1]**

13.3 Figure 13.1 shows the pedigree of a family where some members are affected by dwarfism.

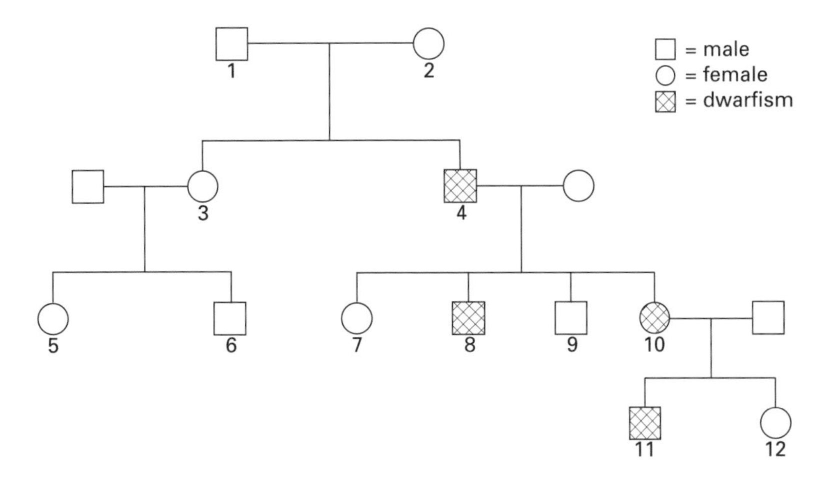

Figure 13.1

In this family the condition arose because of a mutation in a gamete which took part in sexual reproduction.

(a) Which individuals might have produced the mutated gamete? **[2]**

(b) Who was the first person to show a dwarf phenotype? **[1]**

(c) Is dwarfism caused by a dominant or a recessive allele? **[1]**

(d) Explain your evidence for your answer to question **(c)**. **[1]**

(e) Draw a labelled genetic diagram to show the genotypes and phenotypes of all possible children that individuals 5 and 8 might produce. Use D and d for your alleles. **[3]**

(f) What is the probability of a child of 5 and 8 having dwarfism? **[1]**

(g) Is dwarfism a sex-linked condition? **[1]**

13.4 Blood groups depend on the presence of antigens on the surface of the erythrocytes. The MN blood group system is controlled by two alleles M and N which are co-dominant.

(a) Define the term co-dominant. **[2]**

(b) Copy and complete the following table that relates to the MN system.

Phenotype	Genotype	Antigen(s) on red blood cells
	MN	
		N
M		

[6]

(c) Draw a genetic diagram to show the predicted genotypes and phenotypes of the children of parents who are both MN. **[3]**

(d) A gene that has three alleles determines the ABO blood groups. Alleles I^A and I^B are co-dominant and are both dominant to I^O. What is the phenotype of individuals with the following genotypes?

(i) $I^O I^A$ **(ii)** $I^A I^B$ **(iii)** $I^B I^B$ **[3]**

(e) Write down all the genotypes for blood groups O, A, B and AB. **[4]**

(f) Why are people with group O called 'universal donors' and those with group AB called 'universal recipients'? **[2]**

(g) From Figure 13.2 showing the pedigree of the blood groups of a family, give the phenotypes of individuals 1–4. **[4]**

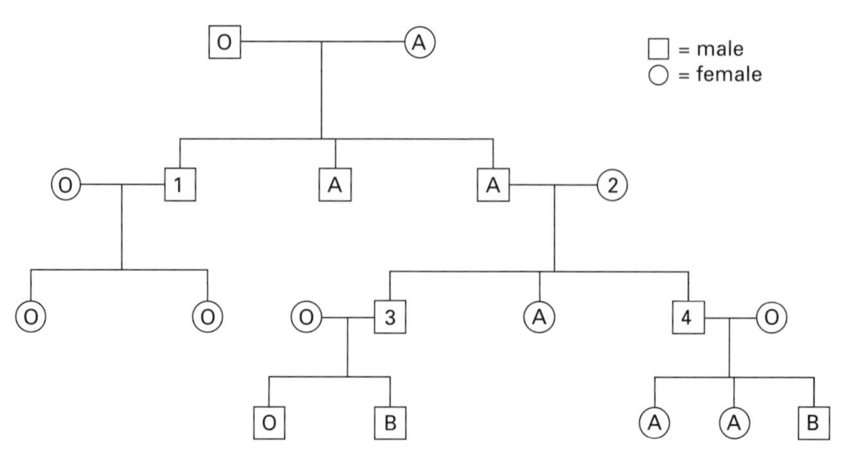

Figure 13.2

(h) A true-breeding red flower and a homozygous white flower were crossed and all the F_1 offspring were pink. The pink flowers were allowed to self fertilise. Their seeds were collected and the following plants grown from them:
259 plants with red flowers, 454 with pink flowers, 217 with white flowers.
Write a short account to explain the genetics of these results. **[8]**

13.5 Tabby markings on cats are determined by two alleles. The agouti allele, A, gives the colour banding and all tabby cats are homozygous or heterozygous for this allele. Then there is a series of alleles for the various tabby patterns, as follows:

T^A – spotted tabby

t^b – blotched tabby

t – mackerel tabby

T^A is dominant to t and to t^b; t is dominant to t^b.

Cats homozygous for t^b are called classic tabby and cats homozygous for T^A are called ticked tabby. (The ticked tabby is always homozygous for the dominant agouti allele.)

(a) Using the tabby allele only, state the possible genotypes of:
(i) mackerel tabby cats
(ii) spotted tabby cats **[4]**

Look at Figure 13.3

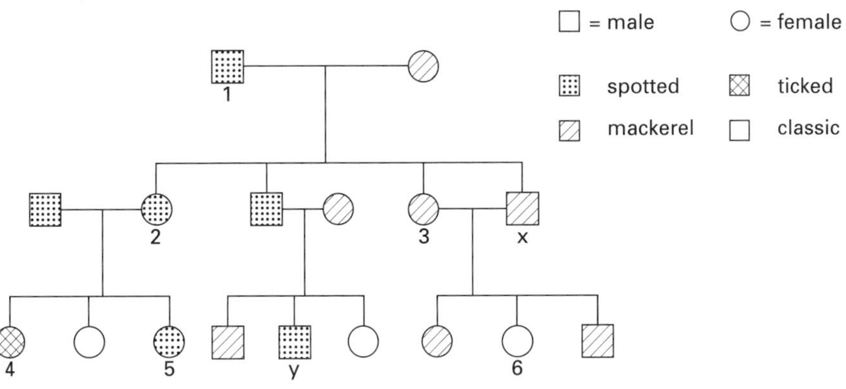

Figure 13.3

(b) Give the genotypes of cats numbered 1–6. **[6]**

(c) Show, using a genetic diagram, the possible genotypes and phenotypes of a mating between cats 1 and 5. **[4]**

(d) Why is it not desirable for cats 3 and X in Figure 13.3 to have mated? **[1]**

(e) A breeder is given the male spotted tabby kitten Y. How can he determine whether it is carrying the mackerel allele? **[4]**

(f) If the breeder found that cat Y is a carrier of the mackerel allele, what kittens might he expect if he crossed it with a homozygous mackerel tabby? Explain your answer. **[3]**

(g) Why is it important in cross **(f)** to be sure that at least one of the parents is homozygous for the dominant agouti allele? **[2]**

13.6 Dihybrid inheritance

(a) What is dihybrid inheritance? **[2]**

The Southern Star plant has two genes carried on separate chromosomes which control flower colour (orange or white) and whether the stem is hairy or smooth. True breeding plants with orange flowers and hairy stems were crossed with white flowering plants with smooth stems. All the F_1 plants had orange flowers and smooth stems.

The dominant allele for flower colour is C and the recessive allele is c.

The dominant allele for stem type is H and the recessive allele is h.

(b) Give the genotype of:

(i) the parent plant with white flowers and a smooth stem

(ii) the parent plant with orange flowers and a hairy stem

(iii) an F_1 plant. **[3]**

(c) Which phenotypic characteristics are recessive? **[2]**

(d) What phenotypes and in what proportions would you expect to get by crossbreeding the F_1 plants? **[2]**

(e) Draw an explanatory genetic diagram to show the outcome of crossing two F_1 plants. Give the genotypes and phenotypes of the F_2 plants. Does your answer confirm your prediction given in question **(d)**? **[7]**

(f) If the H/h and C/c alleles had been linked on a single chromosome, what difference in the proportion of phenotypes would you expect to find amongst the F_2 plants? **[2]**

(g) How would you carry out a cross between the two original parent plants during a breeding experiment? **[4]**

13.7 The table is about dominant and recessive alleles.

Tick the box for the statement which is true and put a cross in the box where it is not.

	Dominant allele(s)	Recessive allele(s)
The probability of heterozygous parents having a child expressing this allele is 1 in 4		
A mother is not affected by this sex-linked allele, but all her sons who inherit it are		
F_1 plants from a cross of different homozygous parents have the phenotype for this allele		
Neither parent is affected by this allele, but some of their children are		
This allele is always expressed in a heterozygous phenotype		
In a dihybrid cross 9/16 of the F_2 generation show these characteristics in their phenotype		
Huntington's disease is an example		
This allele is for colour blindness		

[8]

13.8 **(a)** Complete the passage about sex-linked inheritance, using suitable words to fill the blanks.

Sex linkage refers to the ________________ carried on the sex chromosomes. In most species the ________ chromosome is much smaller than the X. Many alleles on the X chromosome will have no corresponding ________________ on Y and will be expressed in the ______________________ of the individual. This means that a ________________ allele on X will show in the phenotype of the ___________ individual. This leads to special patterns of inheritance for ________________ alleles.

For most animals the female is XX (the _________________ sex) and the male XY. However in birds and butterflies this situation is ________________ . Some insects have X chromosomes only, and so the female is ___________ and the male XO. **[10]**

(b) In the fruit fly *Drosophila melagonaster* eye colour and wing shape are both sex-linked characteristics. Female flies are XX and males XY. The allele for red eye (R) is dominant to white eye (r). The allele for normal wing (N) is dominant to miniature wing (n).

What is the phenotype of the following flies?

(i) $X^N X^n$

(ii) $X^r Y$

(iii) $X^R X^R$

(iv) $X^N Y$

[4]

(c) List the genotypes for the following fruit flies:
  (i) red eyed male
  (ii) female with normal wings
  (iii) white eyed female
  (iv) male with miniature wings. **[4]**

(d) Work out the predicted offspring from the following crosses:
  (i) red eyed heterozygous female × white eyed male
  (ii) female with miniature wings × male with normal wings **[8]**

Haemophilia is a sex-linked disorder in humans where there is an abnormality in the blood clotting factors. Allele H for normal clotting is dominant to allele h, abnormal clotting.

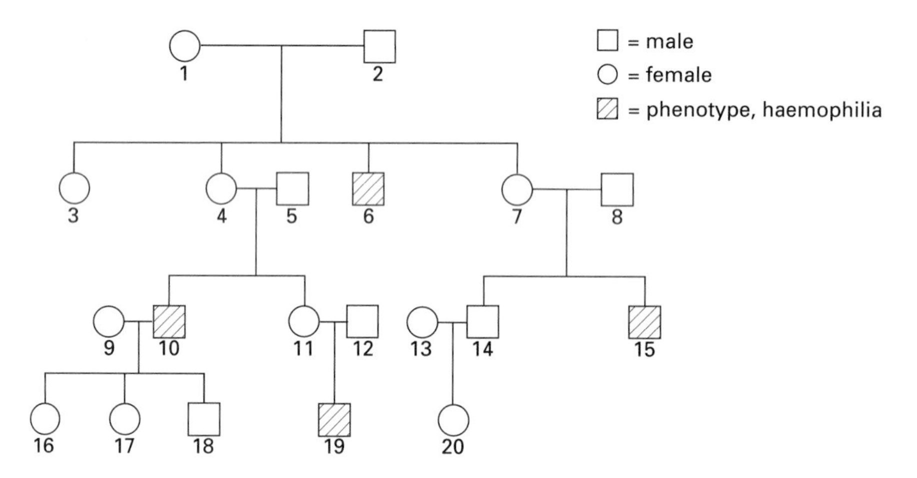

Figure 13.4

(e) From the pedigree shown in Figure 13.4, a family with a history of haemophilia:
  (i) List the genotypes of individuals 1, 7, 8, 10, 11. **[5]**
  (ii) Which individuals in the family are definitely carriers of allele h? **[3]**

(f) Comment on the pattern of inheritance of the H/h allele. **[2]**

13.9 Two unlinked gene loci interact to give variation in coat colour in certain breeds of cat. The dominant allele for colour B produces black pigment; the recessive allele b^1 gives a cinnamon coat. Allele D prevents dilution of coat colour pigment which occurs in animals homozygous for the recessive allele d. Dilution of the black gives a blue coat and dilution of cinnamon produces fawn.

Work out the expected probability, phenotypes and genotypes of the offspring from a cross between a blue cat (Bb^1dd) and a cinnamon cat (b^1b^1Dd). **[8]**

13.10 **(a)** What is a karyotype? **[1]**

Figure 13.5 shows the chromosomes from a cell of a normal human fetus obtained by amniocentesis.

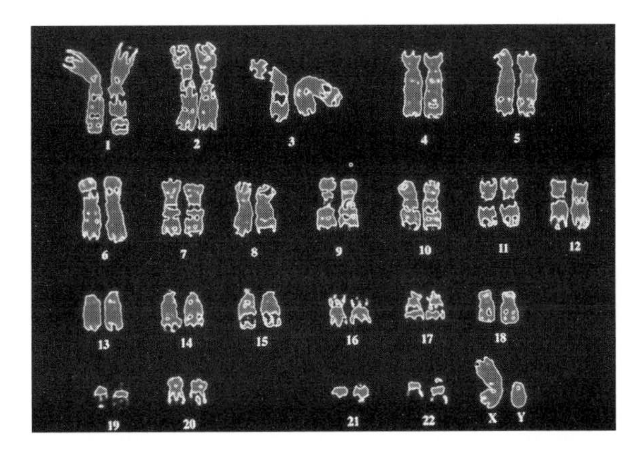

Figure 13.5

(b) How many chromosomes are shown? **[1]**

(c) What sex is this fetus? **[1]**

(d) State two differences that you would see in a cell from a female fetus with Down's syndrome. **[2]**

(e) Describe the genotype and briefly the phenotype of individuals with

(i) Klinefelter's syndrome

(ii) Turner's syndrome. **[4]**

14 Evolution

There is variation between all members of every species due to influence of the environment on the genotype of the individual. Characterisitics that are controlled by polygenes vary continuously; discontinuous variation is controlled by a single gene. Mutations are a source of variation.

The frequency of alleles in a population can be predicted by the Hardy Weinberg equation. Selection pressures – stabilising, directional or disruptive – act on the genotypes in the population. Those individuals best fitted to prevailing environmental conditions will reproduce, and this increases the frequency of their alleles.

New species arise naturally as a result of geographical and reproductive isolation. Artificial selection is important in the developing of new breeds of animals and crops in agriculture.

Hominids have evolved from primates. The earliest remains are believed to have dated from around 4 million years ago. Neolithic man developed the earliest agricultural settlements about 10 000 years ago.

14.1 The height of 200 adult male humans is shown in the table.

Height (m)	Frequency
1.490–1.529	2
1.530–1.569	4
1.570–1.609	9
1.610–1.649	14
1.650–1.689	23
1.690–1.729	32
1.730–1.769	36
1.770–1.809	29
1.810–1.849	23
1.850–1.889	17
1.890–1.929	9
1.930–1.969	2

(a) Present the data in the table in a graph. **[1]**

(b) (i) What name is given to a graph that has this shape? **[1]**
(ii) Explain briefly what the graph shows. **[2]**
(iii) Define the terms mode, median, mean, standard deviation. **[4]**

(c) Complete the following passage by filling the gaps with appropriate words.

A characteristic such as eye colour shows considerable ______________ within the population. This range in __________________ is due to the characteristic being controlled by ______________, that is a number of genes working together. Their effect is described as being _________________, since each ______________ allele contributes to the characteristic. ________________, ____________________ and _________________ are three characteristics of humans inherited in this way. The variation in phenotype may be plotted graphically and providing that the quantitative data is from a _________________________sample, will show a ______________ distribution.

For other characteristics, individuals can be divided into _______________ groups. These features are controlled by _____________________. They are said to vary _________________. Examples in humans are the ABO blood group and _________________. **[14]**

14.2 **(a)** Discuss in a short account of about 250 words, the extent to which the environment has an effect on inherited characteristics. **[6]**

(b) Read the following and for each example explain briefly the extent to which you think environmental and genetic factors affect the phenotype.

A Locusts reared in isolation and then set free live a solitary existence: locusts reared in high density groups and then set free, live and migrate as a swarm.

B There is a slightly higher incidence of left-handedness amongst twins, their parents and their siblings, than is found in families without twins.

C Some rabbit breeds carry the 'himalayan' gene for developing colour points on their ears, nose, paws and tail. In temperate climates the points are a darker shade in autumn and winter than during the spring and summer.

D Amphibious bistort is a plant which grows either in water or on land. When aquatic it has floating elliptical leaves and medium to long petioles (leaf stalks). The plants growing on land have narrow, upright, hairy leaves and short petioles. If cuttings are taken of each type and then grown in the habitat of the other, the new plants always develop the features for the habitat that they have been grown in.

E There is a slightly higher, but statistically significant incidence of left-handed twins amongst twins delivered second. **[10]**

14.3 Gene or point mutations

(a) Define the terms mutation, mutant and mutagen. **[3]**

(b) List five causes of mutation. **[3]**

(c) This is a sequence of codons along a length of mRNA that code for the amino acids in a polypeptide chain.

– UCA – UAU – AAG – GCA – GGU – UUG –

The table overleaf give the codes for some amino acids

Codon	Amino acid	Codon	Amino acid	Codon	Amino acid
UAU	tyrosine	UUG	tryptophan	AAG	lysine
UCA	serine	GCA	alanine	AGG	arginine
UGA	stop code	GGU	glycine	AAA	lysine
UUA	leucine	GUU	valine	CAG	glutamine

(i) Draw the primary structure of this peptide chain. **[1]**

(ii) A mutation occurred at the third codon (from the left) so that guanine was replaced by adenine. Name this type of mutation and comment on the effect the mutation would have on the peptide chain. **[2]**

(iii) The adenine base was deleted from the second codon from the left. Re-write the base sequence following this mutation and explain the consequences of this event in terms of cell metabolism. **[3]**

(iv) Re-write the base sequence following the insertion of a cytosine base into the third position of codon 1. **[1]**

(d) Each step in the following metabolic pathway is catalysed by a single enzyme (1, 2 or 3.)

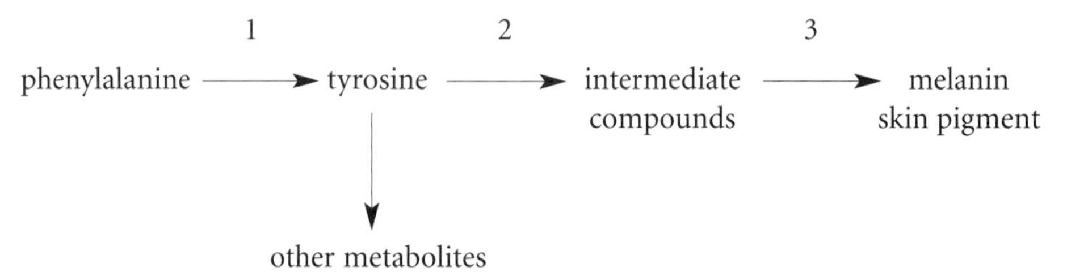

(i) What is the chemical nature of an enzyme? **[1]**

(ii) If a person is homozygous for a recessive mutation that prevents the production of enzyme 1, what would be the effect on this pathway? **[2]**

(iii) Name the condition that occurs in animals that have a mutation that prevents the production of enzyme 3. **[1]**

14.4 Sickle cell

(a) Describe the mutation that occurred to bring about haemoglobin S. **[3]**

(b) What is the difference between sickle cell disease and sickle cell trait? **[3]**

(c) Explain what is meant by the 'evolutionary advantage' of having sickle cell trait, in relation to populations living in areas of the world where malaria is endemic? **[4]**

14.5 Chromosome mutations

(a) Explain what is meant by:

(i) aneuploidy

(ii) polyploidy

(iii) trisomy **[3]**

(b) The table shows data collected from chromosome investigations of spontaneously aborted human fetuses.

	Genetic fault	Frequency
1	trisomy chromosome 8	0.333
2	trisomy chromosome 16	0.333
3	trisomy chromosome 18	0.025
4	trisomy chromosome 21	0.005
5	monosomy sex chromosome (XO)	0.05
6	trisomy sex chromosome	not found

(i) How many categories of fetus show a normal karyotype? **[1]**
(ii) What would be the chromosome number for a fetus in group 1? **[1]**
(iii) Suggest why there was no data for category 6. **[1]**
(iv) What is the chromosome number for a fetus in category 5? **[1]**
(v) What is the frequency of Down's syndrome amongst the fetuses in this study? **[1]**
(vi) What would be the chromosome numbers of gametes producing a Down's fetus? **[1]**
(vii) Explain the term non-disjunction. **[2]**
(viii) Suggest why non-disjunction is understood to occur more frequently in the gametes of females than of males. **[2]**

(c) Approximately 1 in 1000 babies is born with Down's Syndrome. The risk changes with the age of the mother as this table shows:

Maternal age (yrs)	Risk of baby with Down's syndrome
25	1 in 1450
30	1 in 850
35	1 in 380
38	1 in 190
40	1 in 110
45	1 in 30

Why are the majority of babies with Down's syndrome born to younger women? **[1]**

14.6 **(a)** Name the types of chromosome mutation represented in Figure 14.1 **[3]**

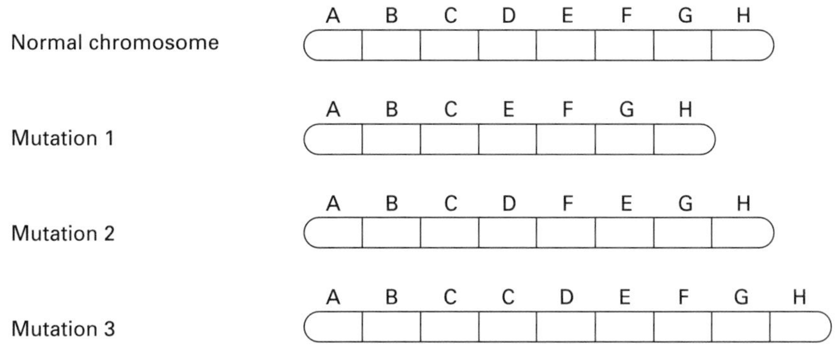

Normal chromosome	A B C D E F G H
Mutation 1	A B C E F G H
Mutation 2	A B C D F E G H
Mutation 3	A B C C D E F G H

Figure 14.1

(b) Describe how the chromosome mutation shown in Figure 14.2 arises. What is it called? **[3]**

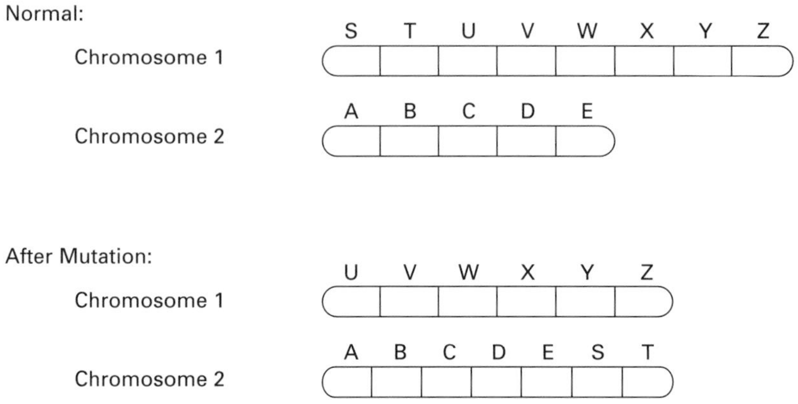

Normal:

Chromosome 1 — S T U V W X Y Z

Chromosome 2 — A B C D E

After Mutation:

Chromosome 1 — U V W X Y Z

Chromosome 2 — A B C D E S T

Figure 14.2

14.7 Hardy Weinberg formula

(a) What is meant by:
(i) a gene pool
(ii) allele (gene) frequency? **[2]**

(b) The Hardy Weinberg formula $p^2+2pq+q^2=1$ is used to predict allele and genotype frequencies in populations.

List five conditions that must be true of a population where the Hardy Weinberg equilibrium applies. **[5]**

(c) In the formula given above what is represented by:
(i) p; **(ii)** q; **(iii)** p^2; **(iv)** q^2; **(v)** $2pq$ **[5]**

(d) What expression is used to calculate the frequency of a single allele? **[1]**

(e) 6300 chromosomes were studied to find the allele for attached ear lobes. 590 such alleles were counted. What is the frequency for the attached ear lobe allele in that population? **[1]**

(f) The agouti gene occurs in mice. Those with genotypes AA or Aa are agouti and those that are aa are non-agouti. A population of mice had 76 non-agouti individuals and 63 that were agouti. What are the frequencies of the A and a allele in that population? **[3]**

(g) The tendency to develop Diabetes mellitus is controlled by a recessive allele. One in 200 people are homozygous for this allele. Calculate the frequency of the heterozygous genotype in the population. Show your working. **[3]**

(h) Data were collected relating to the frequency of the MN blood groups in a population. Alleles M and N are co-dominant. The frequency of the NN phenotype was 0.21. Work out the percentage of the population which have MM and MN phenotypes, assuming that the Hardy Weinberg equilibrium applies. **[3]**

(i) Do the conditions required for the Hardy Weinberg equilibrium usually apply to populations? Explain your answer. **[3]**

(j) What will happen to a population where the conditions for the Hardy Weinberg equilibrium are not met? **[1]**

14.8 Polymorphism

(a) What is polymorphism? **[1]**

(b) *Biston betularia*, the peppered moth, is light coloured and mottled. In 1848 a dark (melanic) mutant form was captured in Manchester. By 1895 98% of these moths in Manchester were the melanic form. The dark colour was due to a mutant dominant gene. The two forms are morphs, the normal form being *Biston betularia typica* and the dark form *B. betularia carbonifera.*

	typica	*carbonifera*
Rural woodland	94.6 %	9.4 %
Industrial woodland	10.1 %	89.9 %

Table 14.1 Observed frequency of the two morphs of *B. betularia*

	typica	*carbonifera*
Rural woodland	13.6 %	86.3 %
Industrial woodland	74.2 %	25.8 %

Table 14.2 Observed frequency of predation of *B. betularia* by woodland birds

(i) Comment on the distribution of the two forms of moth as shown in Table 14.1. **[3]**

(ii) How do the data in Table 14.2 support the idea of Natural Selection? **[4]**

(iii) The data shown in Table 14.1 were collected during the 1950s. Would you predict similar figures if the investigation were to be repeated in the year 2000? **[2]**

14.9 Selection pressures

(a) Name the selection pressure in operation in the graphs shown in Figure 14.3 **[3]**

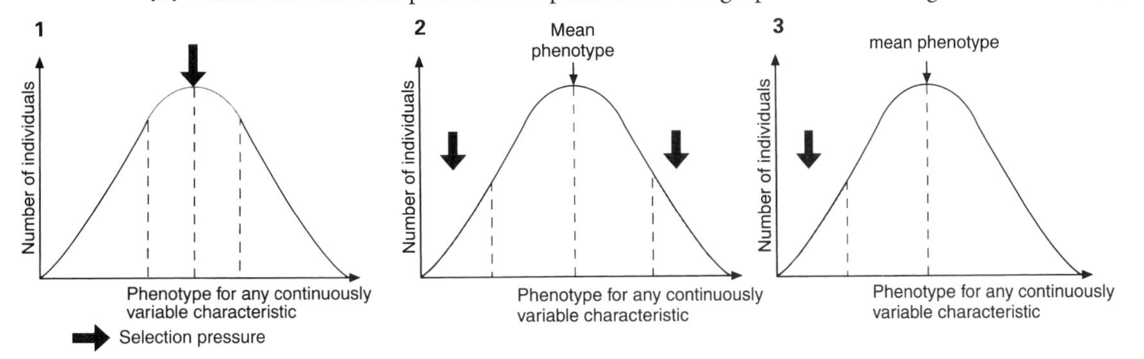

Figure 14.3

(b) In regions of Africa where malaria is endemic which type of selection pressure favours individuals with sickle cell trait? **[1]**

(c) The following phrases refer to the three types of selection pressure that can operate on a population. For each one decide whether it applies to directional, stabilising or disruptive selection.

A the most likely selection pressure to bring about evolutionary change
B eliminates extreme phenotypes from the population
C is the basis for artificial selection
D reduces variation within a population
E fluctuating environmental conditions favour more than one phenotype within the population
F the rarest selection pressure
G operates when there is a gradual change in environmental conditions
H does not promote evolutionary change
I population becomes split into two subpopulations
J the mean phenotype changes in response to favourable environmental conditions
K in operation when environmental conditions are optimal for the phenotype **[11]**

(d) Write a short account to explain why artificial selection is important to humans. **[8]**

14.10 Speciation

(a) Complete the following passage by filling the blanks with suitable words:

Speciation is the process by which one or more new ____________ arise from existing species.
When reproduction occurs between different species the process is known as interspecific ______________ . It is unusual in animals for ____________ hybrids to result from this. Often they are ____________, that is they fail to develop to maturity, or they are ______________ . The mule is such a hybrid, from the mating of a horse and a ______________ .
Reproduction between members of the same species is described as ______________ .
If it occurs in populations which are geographically isolated from one another over long periods of time, ___________________ speciation may occur. Populations of a species may live in the same location but be _________________ by barriers which prevent sexual reproduction. Mechanisms that may prevent breeding include differences in _______________ such as courtship rituals, and _______________ differences. When new species evolve under such conditions ________________ speciation has occurred. **[12]**

(b) The following table compares sympatric and allopatric speciation. Put a tick in the appropriate column if the phrase applies, and a cross if it does not.

	Sympatric speciation	Allopatric speciation
The most common type of speciation in animals		
Hybrids may result		
Occurs in populations living in the same geographical area		
The new species may be polyploid		
Breeding is between members of the same species		

[5]

14.11 Human evolution

Write a short paragraph about each of the following:

(a) The advantages of bipedalism. **[4]**

(b) The importance of discovering fire. **[4]**

(c) Fossils as evidence of human evolution. **[4]**

14.12 Copy and complete the following table.

	Australopithecus	Homo habilis	Homo erectus
Tools	possibly sticks to dig up insect nests		
Food collection		scavengers and gatherers	
Diet			berries, fruits, roots; meat
Use of fire			

[4]

14.13

(a) List three possible reasons for the extinction of Neanderthal Man. **[3]**

(b) Figure 14.4 shows a cave painting discovered in France, from the Upper Palaeolthic era. Besides animals, what other subjects were often used for cave paintings? **[1]**

(c) What technological developments and skills enabled the decoration of deep caves in the Upper Palaeolithic era? **[5]**

(d) What may have been the purpose of a decorated cave? **[2]**

(e) What materials were used by Cro-Magnon Man to fashion decorative objects? **[3]**

Figure 14.4 A cave painting discovered in France

14.14 During the Neolithic era humans became farmers. Give a brief account of the advances that resulted from this. **[8]**

15 Photosynthesis

Photosynthesis is the synthesis of organic compounds by green plants (producers), using simple raw materials – carbon dioxide and water and light energy from the sun. The efficiency of photosynthesis varies depending on light intensity and wavelength, carbon dioxide concentration in the immediate environment, and temperature. A by-product of photosynthesis is oxygen and this replaces losses from the ecosystem due to respiration and the burning of fossil fuels.

The productivity of plants is their gross primary production. Plants are the first stage in food chains by which energy is transferred through the ascending trophic levels. Food chains link to form food webs. The relationships between organisms in a food chain can be represented as pyramids of number, biomass or energy.

15.1 **(a)** **(i)** Which organs in a plant may contain chloroplasts? **[1]**

(ii) Name two tissues in a leaf that are photosynthetic. **[2]**

(b) These are two chloroplasts from a plant cell. Identify labels A–D. **[4]**

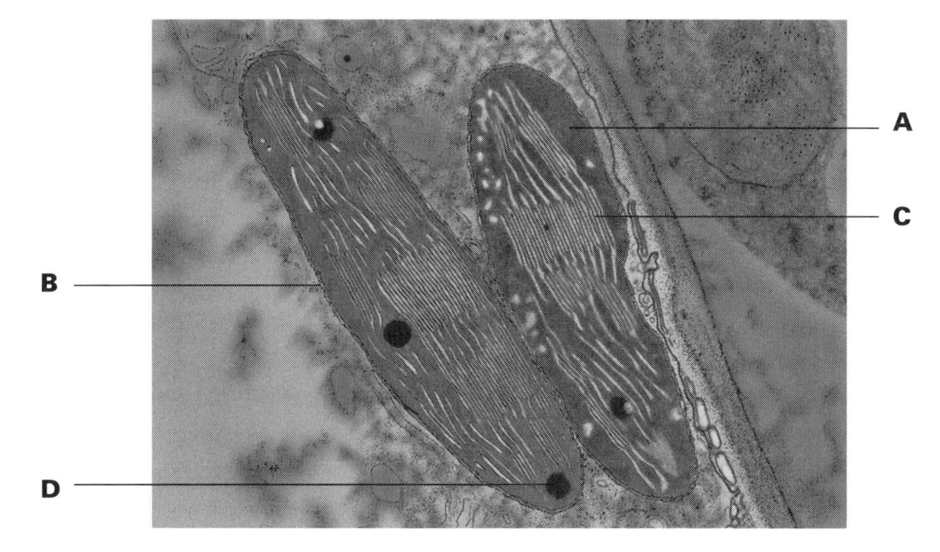

Figure 15.1

(c) Which labels show

(i) where the light independent reactions occur? **[1]**

(ii) where the light dependent reactions occur? **[1]**

(d) These chloroplasts are magnified $\times$ 22 000.
Calculate the actual length of the chloroplast on the left. Show your working. **[3]**

15.2 **(a)** List three reasons why photosynthesis is vital to humans. **[3]**

(b) Write a word and chemical (symbol) equation to summarise the process of photosynthesis. **[2]**

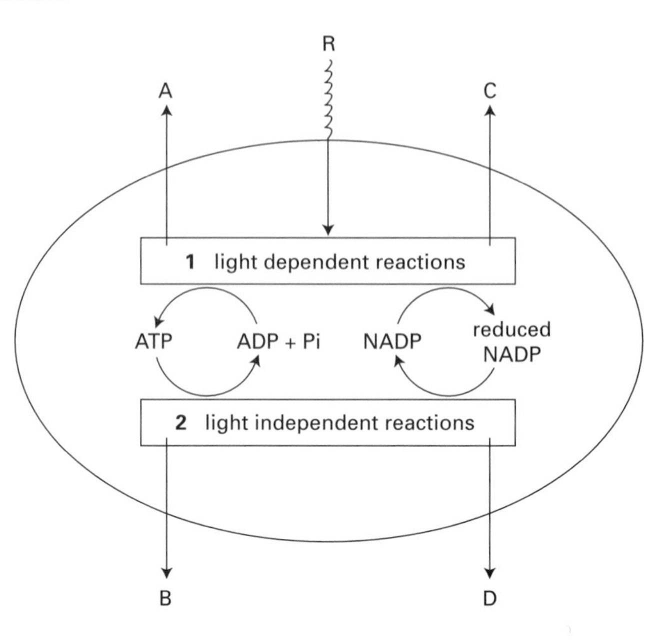

Figure 15.2

(c) **(i)** Exactly where in a chloroplast do stages **1** and **2** occur? **[2]**

(ii) What is energy source R? **[1]**

(iii) Identify compounds A B C D. **[4]**

(iv) Explain whether photosynthesis is an anabolic or a catabolic process. **[2]**

15.3 **(a)** Which wavelengths of light do green plants use most during photosynthesis? **[2]**

(b) Why do most leaves appear green? **[1]**

(c) Write a short illustrated explanation of the terms absorption spectrum and action spectrum for a pigment such as chlorophyll a. **[5]**

(d) Complete the following passage by adding an appropriate word to fill each blank:

Leaves contain a mixture of different _______________ which can be _____________ using the technique of _______________ .
Spots of mixture are put on to an absorbent surface (such as ______________) which is dipped into a suitable _______________ . The components of the mixture differ in their ______________ in the solvent and move at different speeds as the solvent soaks through the medium.
Each ________________can be identified by a unique value found by measuring the distance it moved during the procedure and dividing this by the distance moved by the _________________ . The value that is calculated is the __________ value. **[9]**

15.4 The light dependent reactions

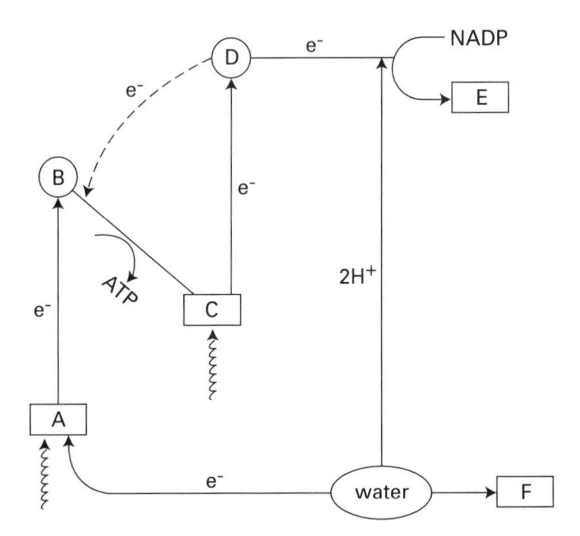

Figure 15.3 Light dependent stage of photosynthesis

(a) List the missing labels A–F **[6]**

(b) What is photophosphorylation? **[1]**

(c) Explain what happens between points B and C. **[3]**

(d) What is the function of NADP? **[2]**

(e) **(i)** What is the name of the process during which water molecules are split? **[1]**

(ii) Write a word and symbol equation to show this process. **[2]**

(f) What products of the light dependent stage of photosynthesis are

(i) released to the environment

(ii) used in the light dependent stage of photosynthesis? **[3]**

15.5 The light independent reactions

(a) Copy and complete the following passage.

During the light independent stage of photosynthesis ____________________ is reduced to ________________ using hydrogen from ____________________ and energy from ________________ . **[4]**

(b)

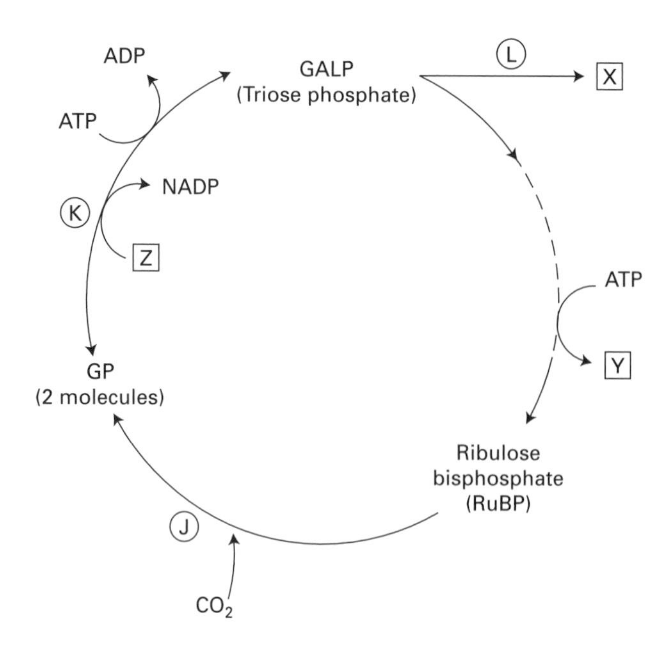

Figure 15.4 Light independent stage of photosynthesis

(i) Identify X Y and Z. **[3]**
(ii) Name the chemical reactions that are taking place at points J, K and L. **[3]**
(iii) What is the function of ribulose bisphosphate (RuBP)? **[1]**

(c) How many carbon atoms are there in one molecule of the following compounds shown in Figure 15.4?
(i) RuBP, **(ii)** CO_2, **(iii)** triose phosphate, **(iv)** compound X? **[4]**

(d) Why is this part of photosynthesis sometimes called the 'Calvin Cycle'? **[2]**

(e) What is the fate of **(i)** NADP, **(ii)** ADP, **(iii)** compound X , shown in Figure 15.4? **[3]**

15.6 Food chains and webs

Look at Figure 15.5.

(a) What is the source of energy for this food web? **[1]**

(b) Explain the term 'organic detritus'. **[1]**

(c) How many trophic levels are shown? **[1]**

(d) What are the producers for this food web? **[2]**

(e) List the primary consumers shown. **[6]**

(f) Which trophic position do the leech and the stone fly nymph occupy? **[1]**

(g) Why is it unusual to have more than four trophic levels in a web such as this? **[2]**

(h) Why is a food web more informative than a food chain? **[2]**

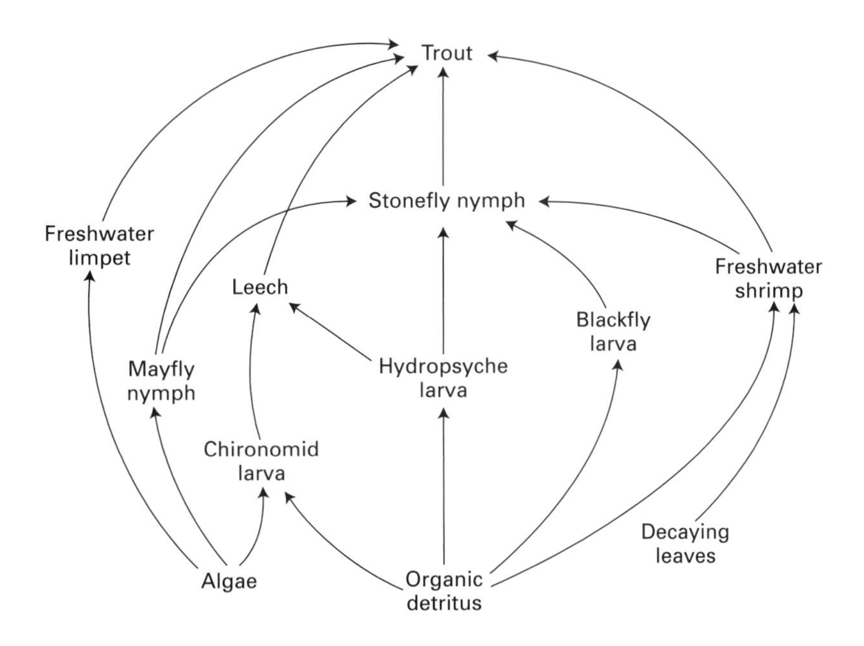

Figure 15.5 Part of a food web in a freshwater stream

15.7 Ecological pyramids

(a) Explain briefly what is meant by **(i)** pyramid of numbers **(ii)** pyramid of biomass **(iii)** pyramid of energy. **[6]**

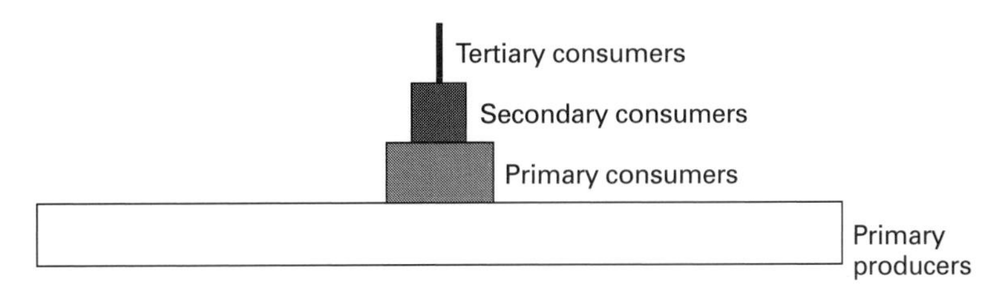

Figure 15.6 A pyramid of numbers for a grassland ecosystem

(b) What information is represented by each horizontal bar in Figure 15.6? **[1]**

(c)

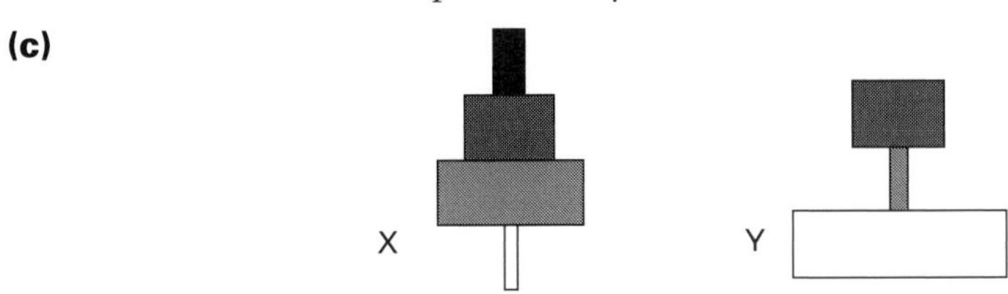

Figure 15.7

(i) Compare the two pyramids of numbers, X and Y, shown in Figure 15.7.

(ii) For each, suggest organisms in a food chain that might produce pyramids of these types. **[4]**

(d) Define the term 'biomass'. **[1]**

(e) Pyramids of biomass may be constructed using data from measurements of either fresh biomass or dry biomass.

(i) Which method is likely to give the most accurate pyramid? Explain your reasoning. **[2]**

(ii) Suggest three difficulties associated with the collecting and interpretation of data for constructing a pyramid of dry biomass. **[3]**

(iii) Comment on the shape of pyramids P and Q. How are the producers in Q able to support the primary consumers in the English Channel? **[3]**

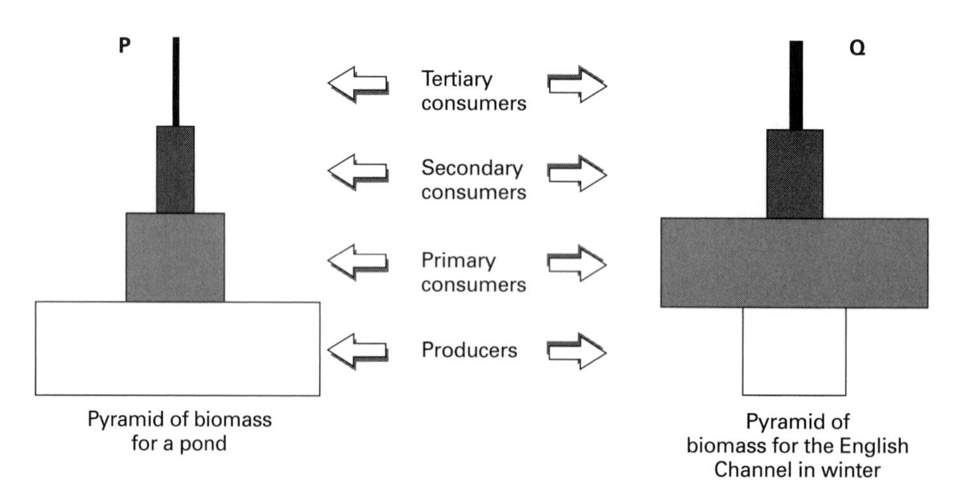

Figure 15.8

(f) (i) Pyramids of energy measure units of energy per area per time. Give one example of SI units that might be used for this. **[1]**

(ii) What is the pattern of energy transfer through the food chain in Figure 15.9? **[1]**

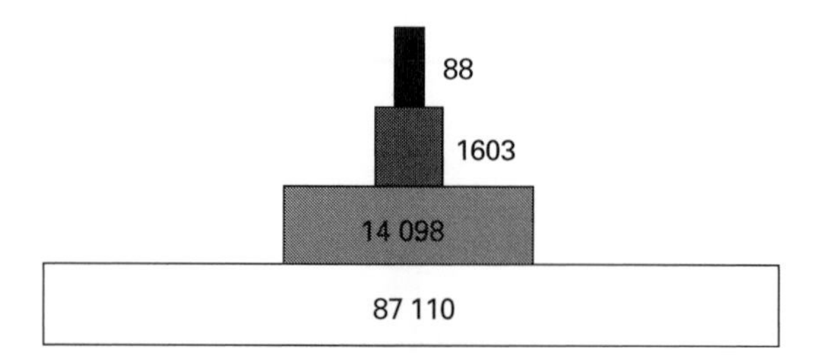

Figure 15.9 A pyramid of energy for Silver Springs, Florida

(iii) Give two reasons why pyramids of energy give a more accurate representation of energetics in a habitat than pyramids of numbers or of biomass. **[2]**

15.8 Energetics

(a) How does a living organism use energy from its food? **[4]**

(b) How is energy transferred away from a food chain at each trophic level? **[3]**

(c) Which groups of living organisms are 'detritivores'? Why are they important? **[3]**

(d) Give two reasons why not all the solar energy falling on a leaf may be used for photosynthesis. **[2]**

(e) Explain the terms gross primary production (GPP); net primary production (NPP); secondary production. **[3]**

(f) The following data show the fate of visible light from the Sun, falling on temperate grassland $m^{-2}\ yr^{-1}$.

	kJ
Energy reflected away from leaves	165 000
Energy used in the evaporation of water from plants	524 000
Energy transferred to the soil	335 000
Primary production	24 950
Energy transferred from plants by respiration	2 950

(i) What is the net primary production of the producers? **[1]**

(ii) Calculate the energy efficiency of primary production as a percentage of visible light falling on the grassland. Show your working. **[3]**

(g) The data show net productivity of four ecosystems.

Ecosystem	Net productivity kJ $m^{-2}\ yr^{-1}$
Temperate woodland	27 000
Temperate grassland (uncultivated)	13 000
Temperate cultivated arable land	24 000
Tropical forest	49 000

(i) List three reasons why productivity is higher in a tropical forest than in a temperate woodland. **[3]**

(ii) Explain why productivity is greater in cultivated land than in uncultivated grassland in the temperate climate. **[3]**

15.9

Write an account of how energy is transferred through a food web. **[10]**

16 Respiration

Respiration is a metabolic pathway that occurs in cells during which an organic substrate is broken down and molecules of ATP are formed. ATP transfers energy to metabolic reactions in cells and is converted to ADP.

Aerobic respiration requires oxygen. The first stage, glycolysis, occurs in the cytoplasm. The substrate is oxidised to pyruvate which diffuses into the mitochondrial matrix. Here it forms acetyl coenzyme A, which enters the Krebs Cycle, a cyclical series of oxidation reactions. Hydrogen removed during the cycle is transferred to hydrogen carriers such as NAD, before entering the electron transfer chain. As electrons from the hydrogen are passed along electron carriers, located on the cristae in the mitochondria, energy is transferred to ADP thus forming ATP. This is called oxidative phosphorylation.

Anaerobic respiration does not use oxygen. In animal cells the end product is lactic acid and in yeast cells the two end products are ethanol and carbon dioxide. The yield of ATP from each molecule of substrate is far lower during anaerobic respiration than in aerobic respiration. Fermentation is an example of anaerobic respiration and has useful applications in industries including brewing and winemaking, yoghurt-making, cheese-making and baking.

16.1 ATP

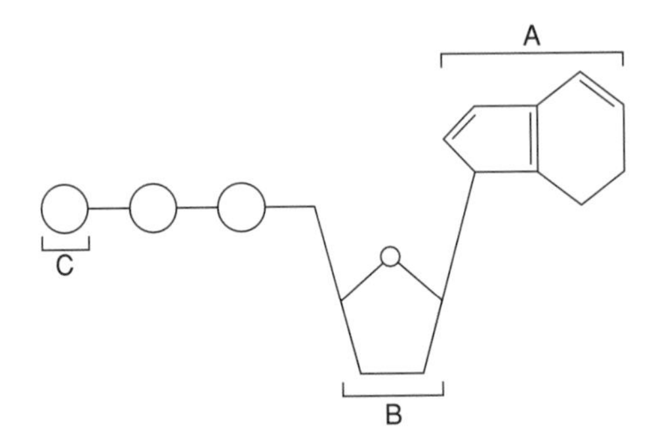

Figure 16.1 A molecule of ATP

(a) What are A B and C? **[3]**

(b) Write an equation to show the conversion of ATP to ADP. What kind of chemical reaction is this? Which enzyme catalyses the reaction? **[3]**

(c) Where in animal cells is most ATP synthesised? **[1]**

(d) What kind of chemical reaction takes place when ATP is synthesised from ADP? Is it exogonic or endergonic? **[2]**

(e) How is ATP produced in a palisade mesophyll cell? **[2]**

(f) Complete the table by inserting a tick in the appropriate column.

	True	False
ATP is the universal source of energy in living organisms		
ATP can be stored in living cells		
ATP is synthesised from ADP by a series of reactions		
Most ATP is synthesised by oxidative phosphorylation in animal cells		
The energy released when a molecule of ATP is converted to ADP is constant		

[5]

(g) State two reasons why ATP is a more useful energy source to a cell than glucose. **[2]**

16.2 Oxidation and reduction

(a) How may a reactant be **(i)** oxidised? **(ii)** reduced? **[6]**

(b) Explain the term 'redox reaction'. **[1]**

(c) For each of the following, say whether the reactant shown in **bold** type is oxidised or reduced during the reaction:

(i) $\mathbf{B} + AH_2 = BH_2 + A$

(ii) $\mathbf{Fe^{2+}} - e^- = Fe^{3+}$

(iii) $\mathbf{Cu^{3+}} + e^- = Cu^{2+}$

(iv) $\mathbf{\frac{1}{2}[O_2]} + H_2 = H_2O$ **[4]**

(d) What name is given to enzymes that catalyse reactions such as **(i)**? **[1]**

16.3 **(a)** Distinguish between cellular respiration and gaseous exchange. **[2]**

(b) Match the following terms to the phrases below:

metabolism catabolism anabolism endergonic exogonic

A a chemical reaction during which a substrate is degraded
B a chemical reaction from which energy is transferred from the reactants
C the sum of the anabolic and catabolic reactions
D the synthesis of a new substrate
E a reaction requiring the input of energy **[5]**

(c) Look at Figure 16.2.

(i) Label W X Y Z. **[4]**

(ii) Where precisely in an eukaryotic cell do each of W X Y and Z take place? **[4]**

(iii) Where in a prokaryotic cell does Z occur? **[1]**

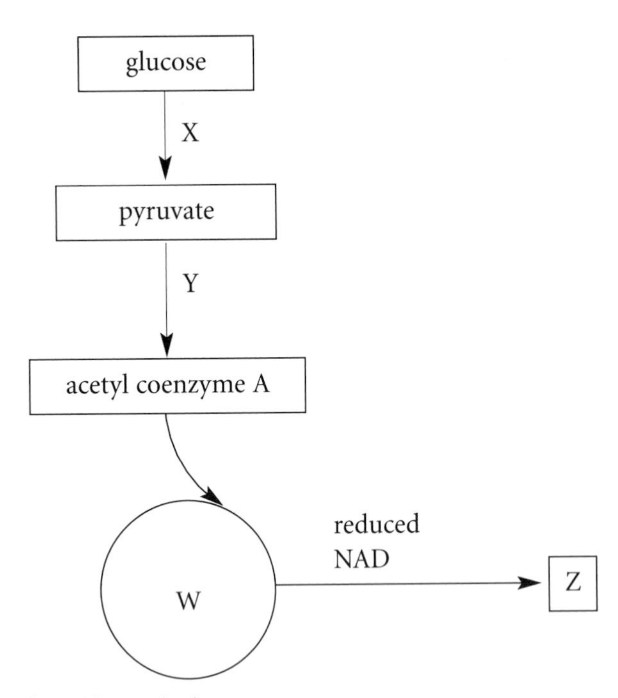

Figure 16.2 A summary of aerobic respiration

16.4 **(a)** Fill the blanks with suitable words, to complete the following passage.

During glycolysis ________________ the most common respiratory substrate, is ______________ using phosphate groups from molecules of ATP. ATP is converted to ____________ . From each molecule of the initial substrate two triose molecules result, having ______________ carbon atoms each. Hydrogen is transferred to ____________ which is reduced. By the end of the pathway there is a net gain of ____________ molecules of ATP.
____________ the end product of glycolysis is converted to ______________ during the ______________ reaction with the loss of ________________ . More ________ is reduced. **[11]**

(b) Write word equations to show what happens to pyruvate under anaerobic conditions in

(i) a yeast cell

(ii) a skeletal muscle fibre **[4]**

(c) Numbers of mitochondria vary in different types of cell. Sperms, cardiac muscle fibres and slow twitch muscle fibres contain many mitochondria; human white blood cells and kidney medulla cells have few mitochondria and red blood cells have none.

Suggest why there are different concentrations of mitochondria in different cells. **[1]**

(d) How is ATP produced in a red blood cell? **[1]**

(e) Why do people who train and exercise regularly have more mitochondria in their muscle cells than people who do not? **[1]**

(f) Many human disorders are recognised which are characterised by abnormal mitochondria (such as small size or abnormal shape). Suggest one effect that this might have on the affected person. **[1]**

16.5 Apparatus Q was set up in a laboratory. The reactants used were ground liver, glucose and ATP. Pure oxygen replaced the air in the apparatus. The flask was kept in a water-bath maintained at 25°C. The level of liquid in the manometer arm was recorded at the start of the experiment and again after 1 hour.

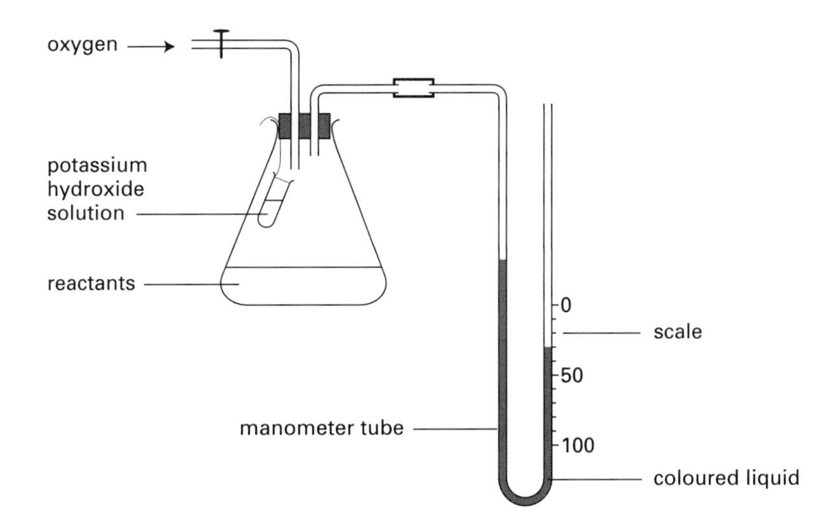

Figure 16.3 Apparatus Q

The experiment was repeated in identical conditions, varying the reactants as shown:

P	R	S
Liver	Liver	Distilled water
Glucose	Distilled Water	Glucose
	ATP	ATP

The following results were obtained:

Flask	Level of liquid in manometer arm	
	At start	After 1 hour
P	10	30
Q	0	90
R	10	20
S	20	20

(a) What is the purpose of the potassium hydroxide solution? **[1]**

(b) What is being measured in this experiment? **[1]**

(c) What is the purpose of Flask S? **[1]**

(d) Comment on the results obtained for flasks P Q and R. **[3]**

(e) How would the rate of reaction in flask Q have been different if the water bath had been at 35°C? **[1]**

16.6 Krebs Cycle

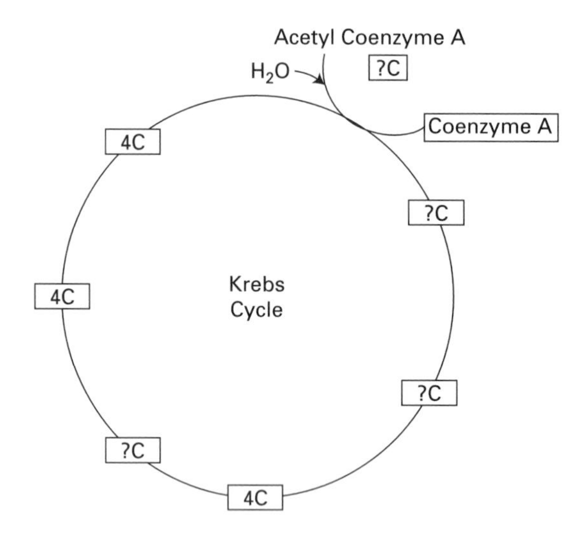

Figure 16.4

(a) Copy Figure 16.4 which shows Krebs Cycle and complete it by adding:
  (i) where carbon dioxide is produced **[2]**
  (ii) where ATP is formed **[1]**
  (iii) where hydrogen carriers have been reduced. **[4]**

(b) Write in the number of carbon atoms that have been omitted (?C). **[3]**

(c) How many turns of Krebs Cycle will there be for each molecule of glucose that is respired aerobically? **[1]**

(d) NAD and FAD are both coenzymes. What is a coenzyme? **[1]**

(e) Which vitamin is NAD derived from? **[1]**

(f) What is the fate of reduced NAD and reduced FAD formed during glycolysis, the link reaction and Krebs Cycle? **[1]**

(g) Name the type of reaction that results in the removal of carbon dioxide during Krebs Cycle. **[1]**

(h) Do the compounds in Krebs Cycle undergo oxidation or reduction? **[1]**

16.7 Electron transfer chain

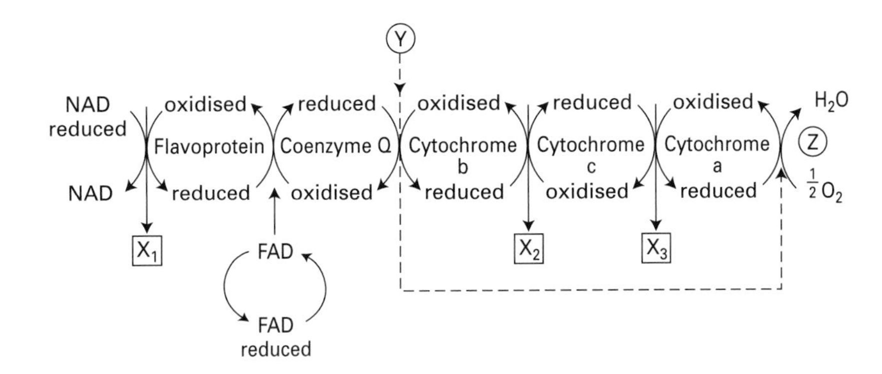

Figure 16.5

(a) Where in the cell is the electron transfer chain found? **[1]**

(b) What passes from reduced NAD and reduced FAD to coenzyme Q? **[1]**

(c) What is produced at points X_1 X_2 and X_3 on the diagram? **[1]**

(d) What happens at point Y? **[3]**

(e) What happens at Z? **[2]**

(f) Use the data in the Redox Potential scale below to explain why the electrons always travel in the same direction along the carrier system. **[2]**

Redox pair	Potential (volts)
NAD reduced/NAD	–32
Cytochrome c oxidised/reduced	+25
Cytochrome a oxidised/reduced	+29
Water/Oxygen	+82

(g) Cyanide is an irreversible inhibitor of cytochrome oxidase. Why is cyanide such an effective poison? **[4]**

16.8 Write a short account of the Chemiosmotic Theory. **[6]**

16.9 **(a)** Glucose is the most common respiratory substrate. Name the compound stored in the liver which is a source of glucose. **[1]**

(b)

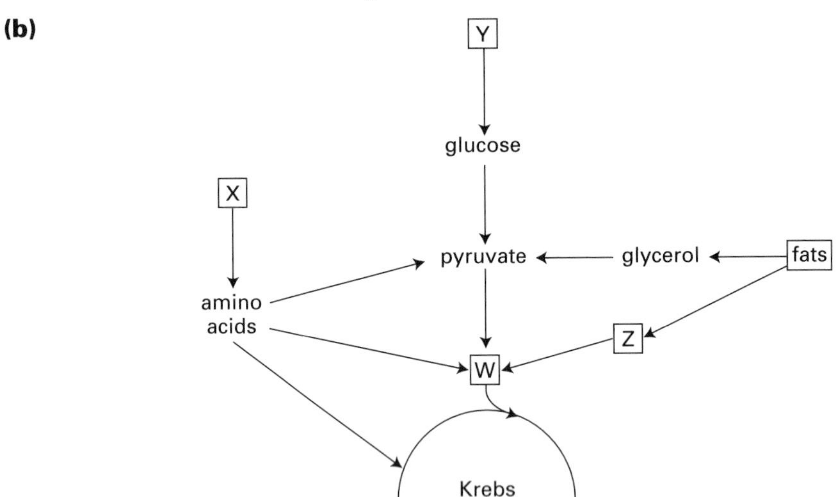

Figure 16.6

(i) Identify W X Y and Z on Figure 16.6. **[4]**

(ii) Under what conditions might X be used in respiration? **[1]**

(c) What expression is used to calculate RQ (respiratory quotient)? **[2]**

(d) The respiration of the triglyceride tristearin is represented by the equation

$$2C_{57}H_{110}O_6 + 163O_2 = 114CO_2 + 110H_2O$$

Calculate the RQ for tristearin showing your working. **[2]**

(e) Explain the following:

(i) If a person is in shock following an accident, their RQ may drop below the normal range of 0.7–1.0.

(ii) A person exercising hard has an RQ above 1.0.

(iii) RQ for a healthy human may vary between 0.7 and 1.0. **[3]**

(f) Is it possible to deduce with certainty from a given RQ value the nature of the substrate being respired by an organism? Explain your answer. **[2]**

16.10 **Anaerobic respiration**

(a) Define the following terms and give an example for each:

(i) obligate anaerobe

(ii) facultative anaerobe. **[4]**

(b) List three products used by humans that are made by a process relying upon anaerobic respiration. **[3]**

(c)

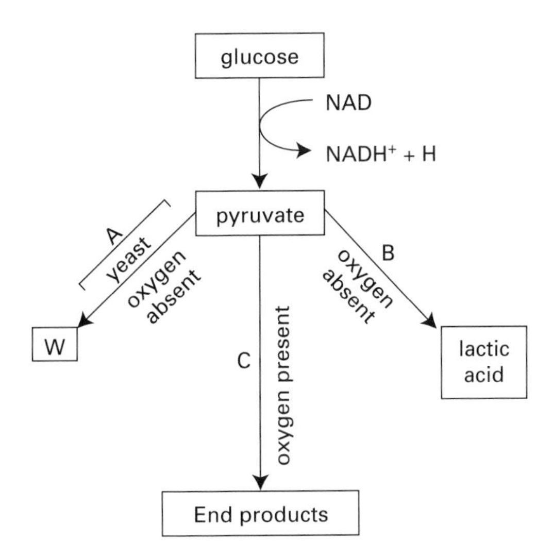

Figure 16.7

(i) How is pyruvate reduced to lactic acid and in which human tissue does this occur? **[2]**

(ii) What term describes stage B? **[1]**

(iii) What is the name of process A? **[1]**

(iv) Identify product W. **[1]**

(v) What are the end products from pathway C? **[3]**

16.11 An experiment was carried out to compare the rate of anaerobic respiration in yeast cells which were cultured in different carbohydrates. $10cm^{-3}$ of yeast suspension in 0.5% carbohydrate solution was drawn into a syringe attached to capillary tubing. The plunger was pushed down to remove air from the syringe. The level of mixture in the capillary tubing was noted and the scale read again after 5 minutes. Any change in the scale was noted in arbitrary units. The apparatus was kept at 30°C throughout the experiment.

Time/min	% substrate remaining		
	Glucose	Sucrose	Lactose
0	100.0	100.0	100.0
5	45.0	85.0	100.0
10	22.5	68.0	100.0
15	15.0	53.0	100.0
20	12.5	36.0	98.5

(a) Draw a graph to show the experimental results. **[4]**

(b) Calculate the rates of respiration for yeast in glucose and sucrose during the first three minutes of the experiment as % substrate used min^{-1}. Show your working. **[4]**

(c) Comment on the activity of yeast with glucose during the second half of the experiment. **[2]**

(d) Suggest an explanation for the curve obtained with lactose. **[2]**

(e) The experiment was repeated using yeast cells that had been cultured with lactose for six hours. Predict how the results might have differed from those shown in the table. **[2]**

16.12 Copy and complete the following table comparing aerobic respiration and photosynthesis.

	Aerobic respiration	**Photosynthesis**
Which kingdoms of living organisms do this?		
Where does this occur in cells?		
When does it occur?		
How is ATP synthesised?	1 2	
What are the raw materials?		
What are the end products?		
Is the process anabolic or catabolic?		

[7]

17 Ecology

Ecology is the study of all aspects of organisms and the environment in which they live. The ability of organisms to compete against one another for food and other essentials, and the abiotic factors prevailing within the ecosystem determine survival. There is constant change as one organism succeeds another.

As we begin the 21st century it is vital that we understand how the rapid industrialisation on the planet over the past two hundred years is bringing about rapid change. Finite resources are being used at an alarming rate. The huge increase in population, pollution from industry and technology, and changes in farming practice are contributing to a situation in which many species are finding it increasingly difficult to survive. Biodiversity is decreasing as many species are lost each year. Conservation is an attempt to save species and conserve genes from extinction.

17.1 **A** Protoctista

B Annelida

C Arthropoda

D Amphibia

E Fungi

Match the following external features to the groups of living organisms **A–E**:

(i) metamorphic segmentation; chaetae; a non-chitinous cuticle

(ii) exoskeleton with a cuticle of chitin; jointed legs and mouth parts

(iii) no chlorophyll; chitin in the cell walls

(iv) single eukaryotic cell

(v) smooth skin; external tympanum. **[5]**

17.2 **(a)** Using examples explain the difference between the following pairs of terms:

biome, biosphere
habitat, ecosystem
population, community
abiotic factor, biotic factor
consumer, decomposer **[10]**

(b) Define the terms **ecological niche** and **species.** **[3]**

(c) Copy and complete the following table for a habitat that you have studied

Habitat:

	Kingdom	Subgroup	Distinguishing external feature
Producer 1			
Producer 2			
Primary consumer			
Secondary consumer			

[13]

17.3 **(a)** Write a short paragraph to explain the term **ecological succession.** **[4]**

(b) Copy the flow chart adding one or more species that might be found at each stage.

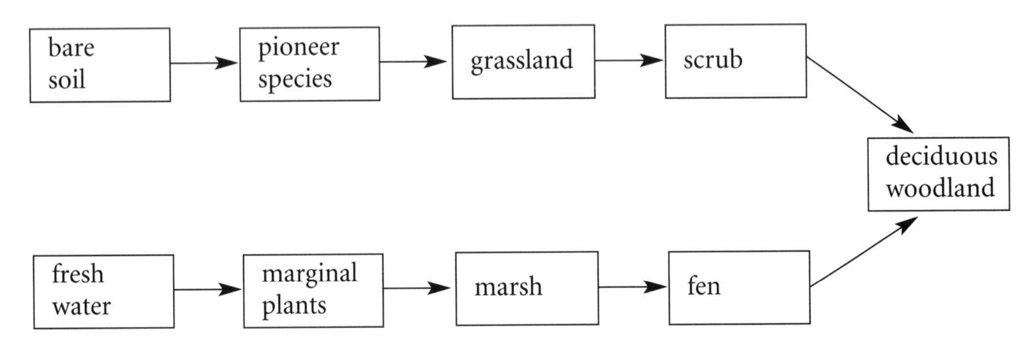

(c) What is meant by climatic climax vegetation? **[1]**

(d) What is the climatic climax vegetation in **(i)** Southern England? **(ii)** Northern Scotland? **[2]**

(e) Study the two examples of succession below

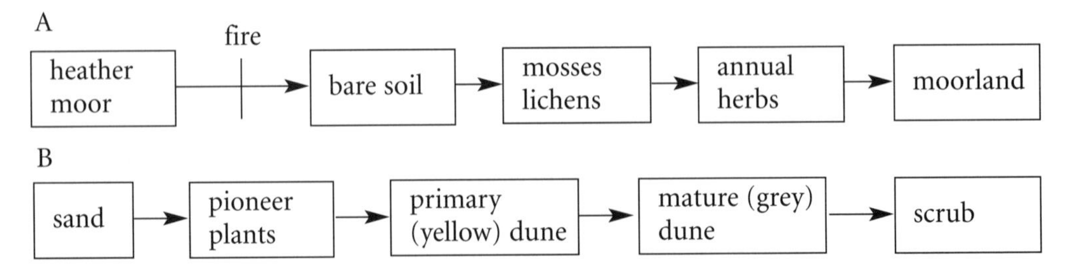

Which shows primary succession and which shows secondary succession?

Explain the difference between the two. **[4]**

17.4 **(a)** Give three sources of organic pollutants of fresh water. **[3]**

(b) Write a short explanation of the events that occur when organic compounds pollute fresh water. Include the terms **biochemical oxygen demand** (BOD) and **eutrophication.** **[5]**

(c) A group of students sampled the fauna of a freshwater stream at two points 200m apart. The total number of animals caught is shown in the table.

	Site A	Site B
Tubificid worm	41	1
Midge larva	23	2
Freshwater louse	7	4
Freshwater shrimp	3	17
Caddis fly larva	0	6
Mayfly larva	0	27
Stonefly larva	0	13

(i) Calculate species diversity for each site using the formula

$$d = \frac{N(N-1)}{\Sigma\, n(n-1)}$$

where N = total number of organisms of all species
n = total number of organisms of each species **[2]**

(ii) Which site showed the greater species diversity? Comment on your answer. **[4]**

(d) Some organisms are known as 'indicator species'. What does this mean? **[1]**

(e) What is suggested by

(i) Large numbers of tubificid worms and low species diversity in a freshwater habitat? **[1]**

(ii) A heavy growth of lichens on roof tops? **[1]**

17.5 **(a)** What is **(i)** a point quadrat, **(ii)** a frame quadrat? **[2]**

(b) **(i)** Explain the importance of random sampling. **[2]**

(ii) If you were instructed to investigate the vegetation of a playing field, what method would you use and how would you ensure a random sample? **[5]**

(iii) Suggest three ways in which you could present your data. **[3]**

(iv) What are the limitations of using quadrats to sample animals? Suggest two animal species that **could** be sampled using a quadrat. **[2]**

(c) The mark-release-capture technique was used to estimate the number of snails *Cepea nemoralis* living in a suburban garden. 54 snails were collected, marked and released. A week later a second collection was made at the same time of day. The total number of snails in the second sample was 61, of which 9 were marked. Calculate the population of *Cepea nemoralis* in the garden using the formula:

$$\frac{\text{number of marked snails in second sample}}{\text{total snails in second sample}} = \frac{\text{number marked snails in population}}{\text{total snail population}}$$ **[2]**

(d) What assumptions are made relating to the behaviour of the snails in this procedure? **[4]**

17.6 A rocky shore is an excellent example of a habitat in which to study the zonation of species.

(a) What method would you use to study the distribution of species from the high water mark for a distance of 150m towards low water? **[4]**

Three species of the brown seaweed *Fucus* show marked zonation on many British shores, the distribution reflecting differing abiotic factors operating in the zones.

Upper shore	*Fucus spiralis*
Middle shore	*Fucus vesiculosus*
Lower shore	*Fucus serratus*

(b) Which kingdom of organisms does *Fucus* belong to? **[1]**

(c) List four abiotic factors faced by *Fucus spiralis* on the upper shore. **[4]**

(d) Many small animals such as limpets, barnacles and mussels are found on the upper shore. How do they depend on *Fucus spiralis* for survival? **[3]**

(e) *Fucus serratus* may be covered by many metres of water at high tide. Other seaweeds in this zone are brown or red. What is the biological significance of their colour? **[3]**

17.7 *Parus caeruleus* (Blue Tit) and *Parus major* (Great Tit) living in deciduous woodland feed their young mainly on caterpillars whose habitat is the local deciduous trees. Eggs are laid between April and June. The graph below shows the mean number of eggs per nest plotted against the date of laying for each species.

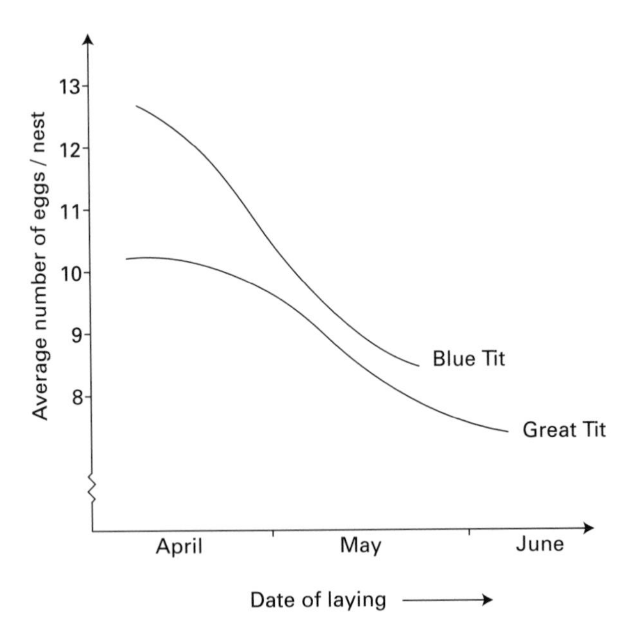

Figure 17.1

(a) Comment on the patterns of behaviour of the two birds. **[3]**

(b) Suggest why the young hatching earlier in the season have an increased rate of survival compared with those hatching later. **[1]**

Eggs laid later in the season hatch 'asynchronously', that is, over a period of a few days.

(c) How would a family of fledglings hatched from eggs laid in early June differ from those hatching in late April? **[2]**

(d) Why do the eggs laid early in the season **not** hatch asynchronously? **[1]**

(e) Suggest two disadvantages faced by fledglings raised later in the season. **[2]**

(f) Give one advantage to parent birds of a late brood. **[1]**

17.8 **(a)** What is meant by competition between living organisms? **[2]**

(b) Distinguish between interspecific and intraspecific competition. **[2]**

(c) Suggest three resources for which animals might compete and four resources for which plants might compete. **[7]**

(d) Population size is affected by density-dependent and density-independent factors.

(i) Explain these terms. **[2]**

(ii) Tick the column that best describes the following:

	Density-dependent	Density-independent
Death of cattle due to a flash flood		
Starvation of some members of a population due to shortage of food		
Rabbits dying during an epidemic of myxomatosis		
Bats dying during winter where there is a shortage of suitable habitats for hibernation		
Total loss of a staple crop due to drought		

[5]

17.9 *Over the past 40 years the range of pest controls available has changed enormously. Almost all the insecticides related to DDT have gone and many organophosphates have been replaced. Now used are a range of alternatives including synthetic pyrethrins developed during the 1970s and 80s containing pyrethrum as an active ingredient. This broad-spectrum insecticide derived from plants is approved for organic use. It kills on contact with the insect pest or when the pest ingests sprayed plant material. In contrast, pirimicab is specific, killing aphids without harming their natural predators, such as ladybirds or bees. It is safe for use where biological control for spider mite and white fly is in operation. It is toxic to melons, courgettes and other cucurbits.*

(a) Why have 'all the insecticides related to DDT' gone out of use? **[3]**

(b) What is a 'broad spectrum insecticide'? **[1]**

(c) Why is pyrethrum licensed for organic use? **[1]**

(d) Give an advantage and a limitation of pirimicab. **[2]**

(e) What is Integrated Pest Management (IPM)? **[2]**

17.10 Compare the advantages and disadvantages of using biological and chemical methods in pest control. **[10]**

17.11 **(a)** Why do plants and animals need nitrogen? **[1]**

(b) Approximately how much of the atmosphere is nitrogen? Why is the gas not of direct use to most organisms? **[3]**

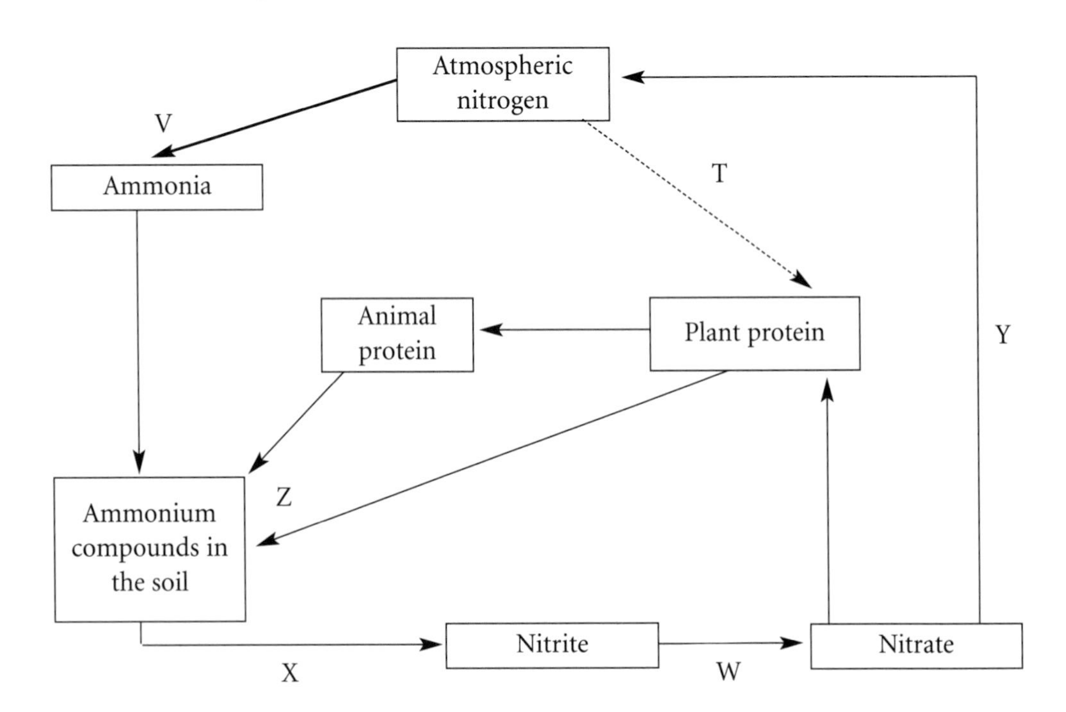

Figure 17.2

(c) Name a genus of bacterium found at each stage V, W and X. **[3]**

(d) What is a chemoautotroph? Give the name of a chemoautotroph that takes part in the nitrogen cycle. **[3]**

(e) What happens at stage Z? **[3]**

(f) Stage T represents the fixation of gaseous nitrogen by *Rhizobium* the bacterium found in the root nodules of plants in the pea family. Explain the relationship between these two organisms. **[3]**

(g) Name a micro-organism active at stage Y. Under what environmental conditions will it be active? **[3]**

(h) Explain one natural process, not shown in Figure 17.2 by which atmospheric nitrogen enters the soil. **[2]**

(i) How can the nitrogen content of soil be increased by cultivation? **[1]**

18 Humans and environment

The global population is increasing rapidly, with far-reaching effects on the need for more resources, such as food. Hence, the impact of human activities on the natural environment is also becoming greater. Major concerns include the use of resources which cannot be replaced, such as rainforest, and pollution caused by wastes we generate.

The best approach appears to be to learn to manage the use of natural resources and disposal of waste in a way which is sustainable.

18.1 Human population

The data in the table and in Figure 18.1 show some of the trends in population for the UK. The map shows the relative population growth in different regions.

Mid-1999 UK population estimates

UK countries	**Population estimate**
United Kingdom	59 500 900
England	49 752 900
Wales	2 937 000
Scotland	5 119 200
Northern Ireland	1 691 800
English region	**Population estimate**
North East	2 581 300
North West	6 880 500
Yorkshire and Humberside	5 047 000
West Midlands	5 335 600
East Midlands	4 191 200
East	5 418 900
London	7 285 000
South East	8 077 600
South West	4 935 700

(a) How do the data support the statement that London and the South East were economic 'hotspots' towards the end of the 20th century? **[2]**

(b) There is a trend towards decreasing household size. Why is this, and what effect does this trend have on the total number of households? **[2]**

(c) Which three areas experienced the slowest growth in population, and what factors might account for this? **[5]**

(d) England has a population density of over 340 people per square kilometre. Describe some of the environmental impacts associated with high population density. **[4]**

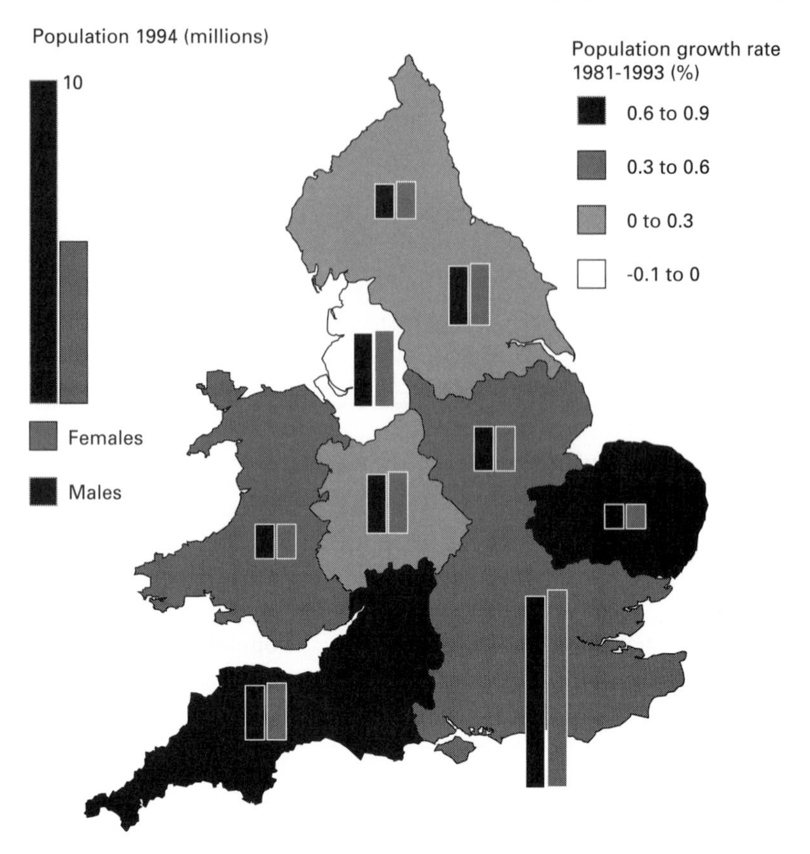

Figure 18.1 Office of Population and Censuses Surveys

18.2 Biodiversity and conservation

(a) Write an account of what is meant by biodiversity, illustrating with reference to different biomes. **[10]**

(b) Why is biodiversity important to the biosphere? **[2]**

(c) What is the main effect of agriculture on biodiversity? **[1]**

(d) Briefly explain why an agricultural practice such as burning stubble affects invertebrate biodiversity. **[2]**

18.3 Malaysia is a rapidly industrialising nation located on the Pacific Rim, with around 70% forest cover (in comparison to under 10% in UK). There are state conservation programmes although less than 10% is undisturbed rainforest and much of the forested land is rubber or oil-palm plantation.

A rainforest is a dynamic ecosystem, often with much local variation and many species represented by only one or two individuals. In Malaysia dipterocarps are huge trees that grow to over 65m which are very valuable commercially. They have a long life cycle, taking 60 years to mature, though the rate of seed production and viability is low. Accordingly, total clearance is rare and the forest is managed by selective logging.

Read the above passage and answer the questions.

(a) Why does the Malaysian government sponsor forest conservation programmes? **[2]**

(b) What is the effect of deforestation on biodiversity? **[1]**

(c) Why does clearance of even a small area of undisturbed forest have a significant effect on biodiversity? **[2]**

(d) How does a forest management method such as selective logging help to maintain the natural structure of the rainforest? **[3]**

(e) What are the main atmospheric effects of deforestation? **[2]**

(f) Why does forest clearance cause soil erosion? **[2]**

(g) Suggest ideas for the types of international cooperation which might help conserve rain forests. **[2]**

18.4 Suggest advantages of each of the following conservation management techniques:

(a) Setting aside land as an Area of Outstanding Natural Beauty **[2]**

(b) Designating an area as a Site of Special Scientific Interest (SSSI) **[2]**

(c) Planting hedgerows **[2]**

(d) Controlling the water table in an estuary **[2]**

(e) Providing signed footpaths in the countryside **[2]**

(f) Setting levels for sustainable fish catches **[2]**

(g) Sustainable forestry **[2]**

(h) Removing bracken from heathland **[2]**

(i) Culling deer in woodland areas **[2]**

(j) Building structures for dune protection **[2]**

18.5 Pollution

(a) Copy and complete the table about causes of water pollution. Give two examples for each source of pollution.

Source of pollution	Examples of pollutant(s) from this source
Industry	
Farming	
Domestic	
Burning fossil fuels	

[8]

(b) How are species of water invertebrates useful in assessing the pollution of a waterway? **[2]**

18.6 Explain what is meant by each of the following terms:

(a) algal bloom **[2]**

(b) eutrophication **[2]**

(c) acidification **[2]**

(d) biochemical oxygen demand **[2]**

(e) indicator species **[2]**

18.7 The table shows relative potency of some greenhouse gases.

Gas	Relative potency as a greenhouse gas	Amount in atmosphere (ppmv)
CO_2	1	365
Methane	10	1.75
Nitrous oxide	270	0.31
CFCs	7000	–

Note: ppmv = parts per million by volume

(a) What is meant by
  (i) global warming? **[1]**
  (ii) greenhouse gas? **[1]**

(b) Explain why CO_2 is such an important greenhouse gas even though its potency is lower than that of other gases. **[1]**

(c) Name one natural source of methane. **[1]**

(d) What was a common use of CFCs in industry until recent years? **[1]**

(e) Describe three predicted environmental effects of global climate change. **[3]**

(f) Why is an international strategy required to tackle an environmental problem such as global warming? **[2]**

18.8 Copy and complete the following passage about the use of fossil fuels, putting appropriate words in the spaces.

Combustion is a process involving __________ . We burn an increasing amount of fossil fuels such as __________, __________ and __________ to provide energy for domestic and industrial power, and for transport. These fuels produce non-metal __________ such as carbon dioxide, __________ and __________ . All these substances are __________ so they lower the pH when dissolved in rainwater, affecting soil mineral balance and plant growth.

Some natural processes such as respiration and __________ also produce carbon dioxide, but the effects tend to be on a __________ scale than those caused by burning fossil fuels. Additionally, there is a balance with other natural processes such as __________ which remove carbon dioxide from the air. **[11]**

18.9 The formulae for some non-metal oxides are shown below. Give the name for each oxide.

(a) CO_2 **[1]**

(b) NO_2 **[1]**

(c) NO **[1]**

(d) CO **[1]**

(e) SO_2 **[1]**

(f) N_2O **[1]**

18.10 Ozone is found naturally in the atmosphere at low levels (the troposphere) and high levels (the stratosphere).

(a) Describe one reason why ozone depletion has occurred in the stratosphere over the last two or three decades, and where there is most evidence of a 'hole' in the ozone layer. **[2]**

(b) What effects does ozone depletion cause? **[2]**

(c) What is the effect of ozone in the troposphere? **[2]**

18.11 Describe some general ideas for reducing the impact of road traffic on air pollution and better use of resources. **[8]**

18.12 Biological control

The graphs show the number of spider mites on cucumbers grown in glasshouses. Pesticide is applied at the points indicated by arrows on the top graph. The lower graph represents the number of spider mites in the batch treated with a natural predator (biological control).

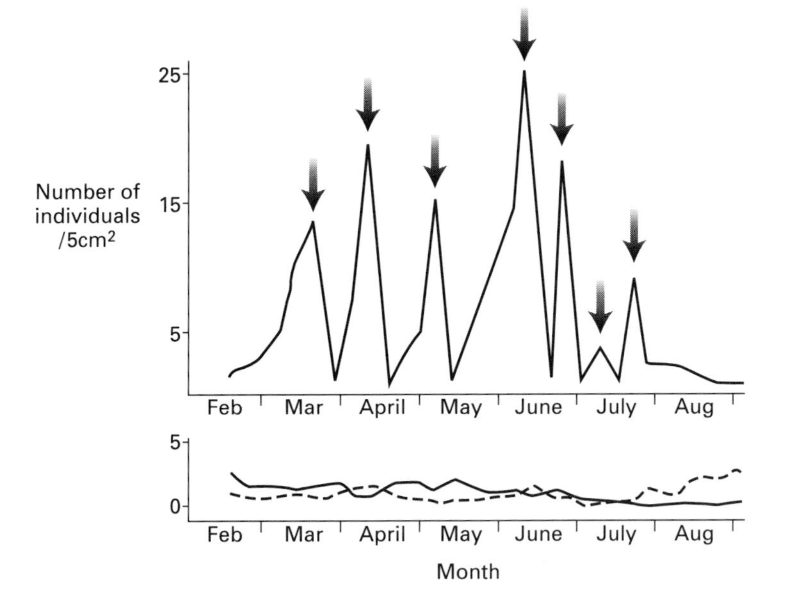

Figure 18.2 Control of spider mite on glasshouse cucumbers

(a) Describe the spider mite infestation rate on cucumbers for those controlled with chemical pesticide. **[3]**

(b) Suggest a reason for the points showing a rapid decrease of spider mite. **[2]**

(c) What is meant by biological control? **[1]**

(d) How does the infestation rate differ for the plot involving biological control? **[2]**

(e) In this case, a species of white fly is a natural predator of red spider mite. The management control method is first to introduce the pest to the glasshouse, followed by the predator (which is bred commercially to ensure continuous supplies).

(i) Suggest three advantages of using biological control. **[3]**

(ii) Suggest reasons why some cucumber growers might prefer to use pesticides rather than biological control, despite its success. **[3]**

18.13 Fishery management

Fishing practices changed during the 1900s to include the use of stronger and lighter plastic nets, radar and sonic detection techniques and on-board refrigeration and freezing facilities. By the mid-century, problems with catch size and an increasing proportion of small fish within the catch were experienced. Towards the end of the 20th century, both sea and freshwater fishery management became an important political and economic issue in Europe. Consequently, fishing quotas, permissible mesh sizes for nets and other key factors were enforced through legislation.

(a) Suggest how modernisation in fishing practices initially affected catch size. **[1]**

(b) What appear to be the longer term effects of modern fishing methods? **[2]**

(c) Why is fishing both an economic and political issue? **[1]**

(d) What would be the initial effects on fishermen of introducing fishing controls? **[2]**

(e) Suggest how the particular fishing management controls mentioned above might improve the situation in the longer term. **[3]**

(f) Suggest one other management control which might help stabilise fish stocks. **[1]**

19 Reproduction

Reproduction is the process which allows the continuation of the human species by producing new offspring. It is the start of the human life cycle which begins at conception and ends with death. Conception involves the fertilisation of gametes (sex cells), produced by specialised sex organs.

In humans embryonic development is internal, lasting around 9 months.

19.1 Sexual reproduction

Explain these terms:

(a) primary sexual organs **[1]**

(b) secondary sexual organs **[1]**

(c) gamete **[1]**

(d) somatic cell **[1]**

(e) haploid **[1]**

(f) zygote **[1]**

19.2 **(a)** What is the role of human reproduction? **[2]**

Table 19.2 *Fertility in different parts of the World.*

World Region	Fertility Rate (births per woman)		% Change in fertility rate 1963–1993
	1963	1993	
Middle East/North Africa	6.9	4.5	34.78
South Asia	6.0	4.1	31.67
Latin America/Caribbean	5.9	3.1	47.46
Industrialised countries	2.8	1.7	39.28
Developing/Semi-industrialised countries	2.7	1.9	29.63

(b) Comment on the data in the table. **[3]**

(c) Suggest four factors which are an important influence on family size. **[4]**

(d) In many parts of the world, selective pressures are not operating on the human population in the same way as they are on other species. What is meant by selective pressure and how does it affect global population trends? **[3]**

19.3

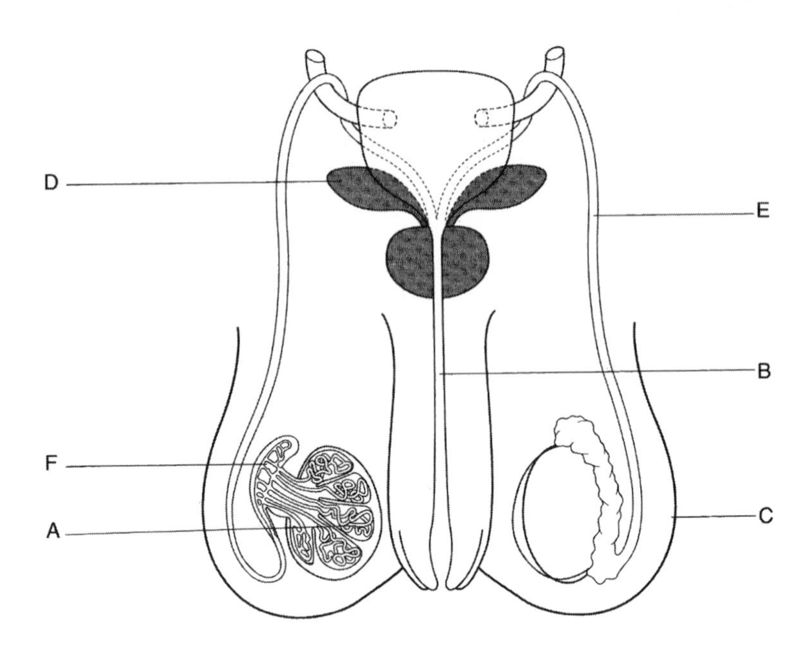

Figure 19.1 Male reproductive system

(a) Name structures A–F and write a sentence to explain the role of each. **[12]**

(b) Copy and complete the table of features of the female reproductive system by putting a tick if a statement about a feature is true and a cross if the statement is false.

Feature	Built of diploid tissue	Produce haploid cells	Implantation occurs here	Fertilisation occurs here	Lost during menstruation
Endometrium					
Cervix					
Ovary					
Vagina					
Oviduct					
Fimbriae					

(c) What is a smear test and why is it offered to women? **[3]**

19.4 **(a)** Gametogenesis is the word used to describe the production of new __________, which in the case of females are __________. Epithelial cells within the __________ which is the outer layer of the ovary, divide by __________, forming __________. All these cells are __________, containing sufficient genetic material to form paired chromosomes. An oogonium may mature to form a __________ which becomes surrounded by cells and is called the primary follicle. All this happens before birth. The remaining changes happen after __________ when follicles begin to mature regularly during the menstrual cycle. Meiosis progresses to anaphase II, so that __________ have a __________ genetic complement. The secondary oocyte is released from the ovary during ovulation. Meiosis completes if __________ occurs, giving rise to an ovum. **[11]**

(b) Suggest labels for structures A–F in Figure 19.2 which shows the ovary at various stages of gametogenesis. **[6]**

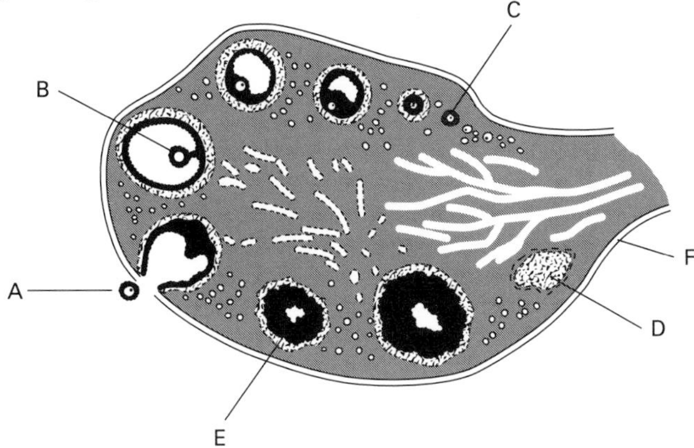

Figure 19.2

(c) Copy and complete Figure 19.3 which shows male gametogenesis, by putting n or 2n in the appropriate spaces (1–6). **[6]**

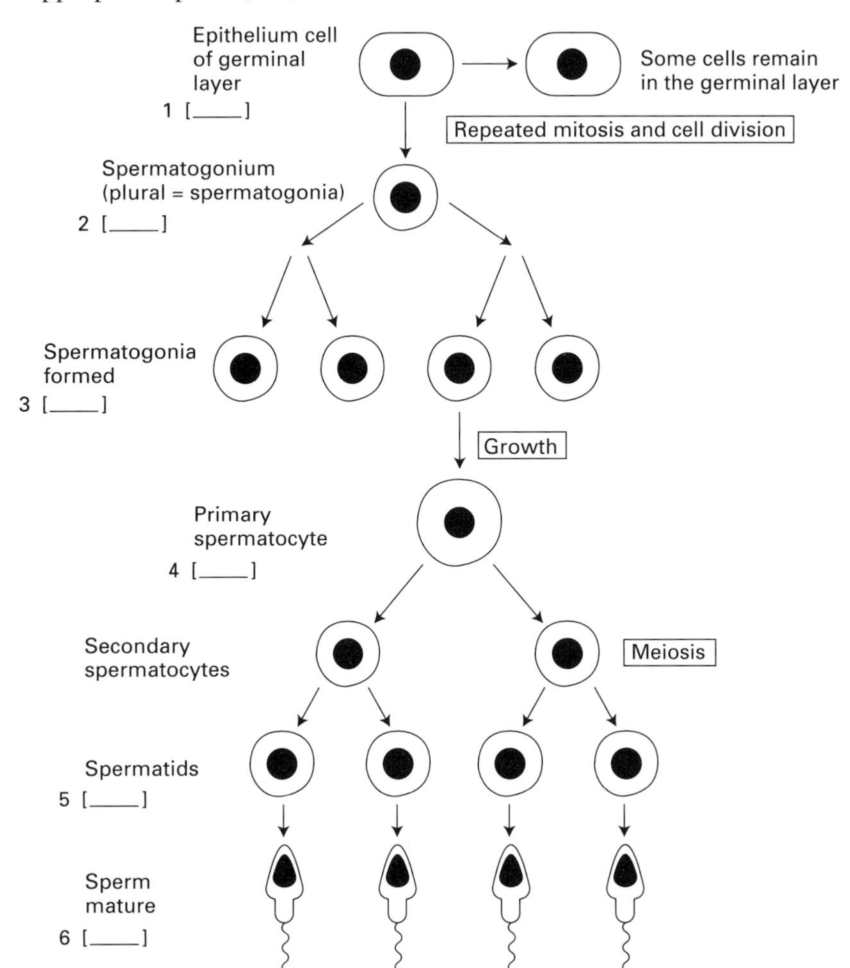

Figure 19.3

19.5 Figure 19.4 is taken from a photomicrograph of a testis. Suggest labels for each of the features A–E. **[5]**

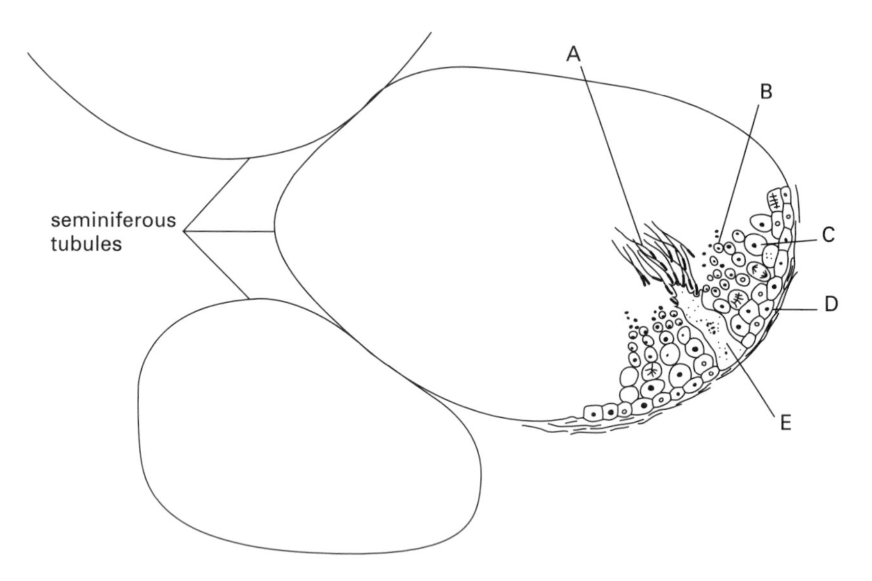

Figure 19.4

19.6 Contraception

Write an account of the main methods of contraception, explaining the biological basis of each. **[10]**

19.7 Conception

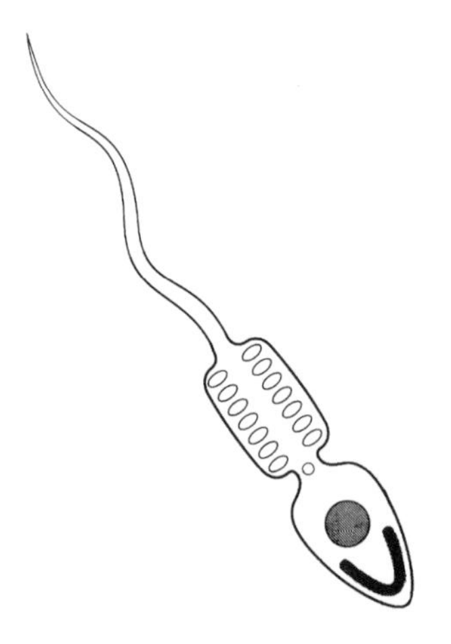

Figure 19.5 Structure of the human sperm

(a) What are the main components of the sperm? **[3]**

(b) What is the role of the acrosome? **[2]**

(c) How are further sperms prevented from entering the oocyte once one sperm has entered? **[2]**

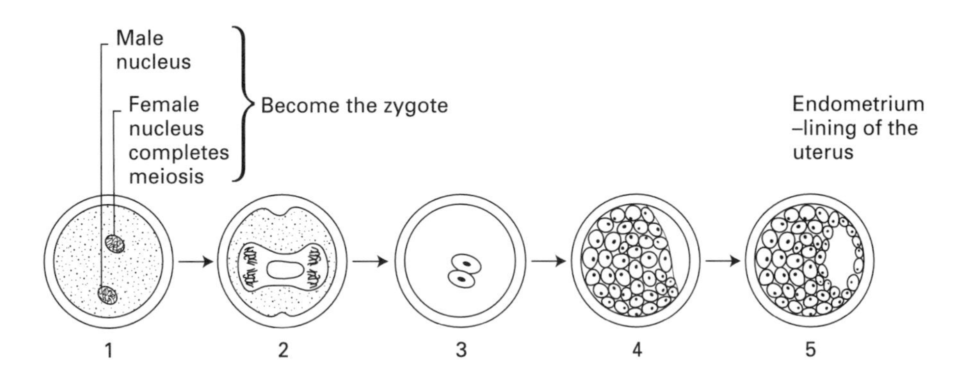

Figure 19.6 Five stages in early embryo development

(d) What immediate change happens to the oocyte after fertilisation? **[1]**

(e) Which type of cell division occurs to increase cell numbers? **[1]**

(f) Name the ball of cells shown in the fourth stage. **[1]**

(g) What is the name of the stage when the blastocyst becomes embedded in the uterus wall? **[1]**

(h) Suggest reasons why a miscarriage may occur. **[3]**

19.8 **(a)** How is the conception of identical twins different from the formation of non-identical twins? **[4]**

(b) Identical twins usually have separate amniotic sacs, but not always. Explain one disadvantage of the twins sharing an amniotic sac. **[2]**

(c) What is meant by dizygotic? **[1]**

19.9 Imagine that a friend wants to talk over his/her worries about starting a family, because of a family history of cystic fibrosis. What issues might you discuss with them concerning genetic screening and counselling, to help them consider their options? **[5]**

19.10 Pregnancy

(a) Copy and complete the table giving information about the functions of the placenta.

Function	Comments
Nutrition	
Respiration	
Excretion	
Protection	
Endocrine	

[10]

(b) What are the chorionic villi? [2]

(c) How are the chorionic villi protected from the maternal immune system? [1]

(d) What tissue forms the selective barrier which separates the fetal and maternal blood supply? [1]

(e) Name two different types of potentially harmful materials which can cross this barrier. [2]

19.11 Birth

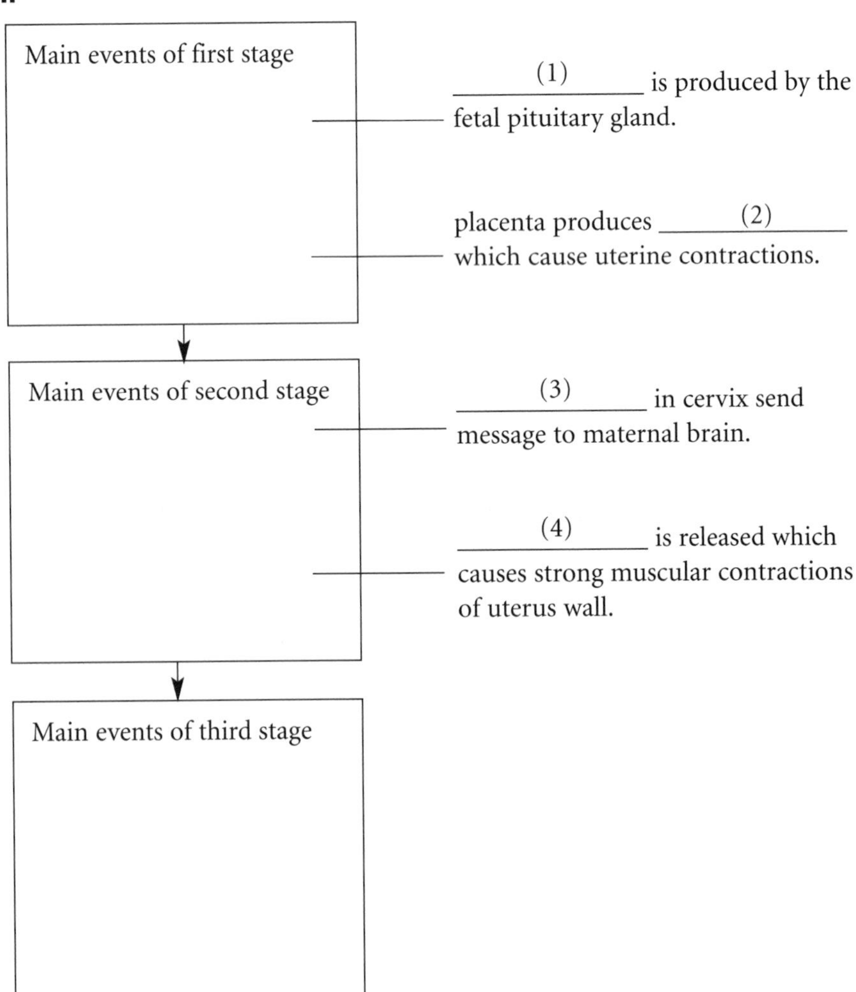

Figure 19.7 Birth

(a) Copy and complete the flow chart of birth stages by writing appropriate phrases to describe the main events happening at each stage, and by filling the blanks in the adjacent text. [10]

(b) One of these hormones is present after birth, affecting muscle fibres within breast epithelial tissue. Why is this hormonal influence important ? [2]

(c) Which other hormone is involved in lactation and what is its role? [2]

19.12 Infant nutrition

(a) Construct a table of pros and cons of breast feeding compared to using formula milk, using these criteria: nutrition, convenience, hygiene considerations, immunity, cost. **[10]**

(b) Comment on the data in Figure 19.8. **[5]**

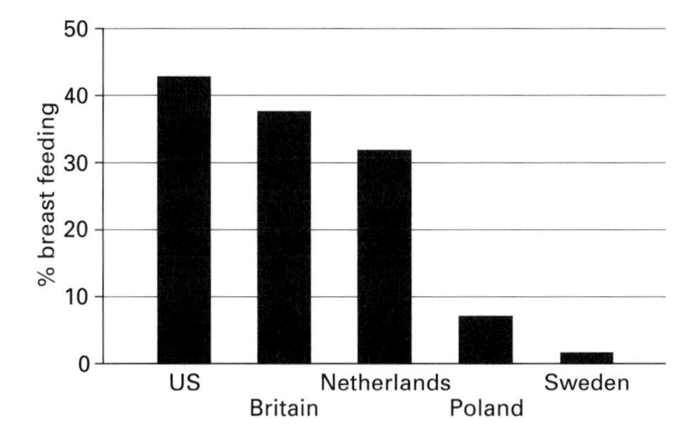

Figure 19.8 Bottle feeding rates for different countries

19.13 Maternal physiology

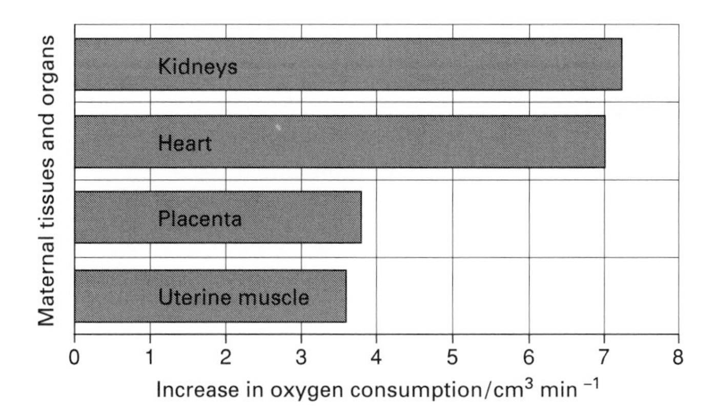

Figure 19.9 Increased oxygen consumption in pregnancy

(a) Explain why there is an increased oxygen demand in each of these tissues. **[4]**

(b) What is the total increase in oxygen consumption shown here? **[2]**

(c) How is this increased demand linked to changes in the mother's cardiac output? **[2]**

(d) Why is there a slight increase in maternal body temperature during pregnancy? **[1]**

20 Life stages

Between conception and death there is a series of life stages from infancy to childhood, and from adolescence to adulthood. Growth mostly occurs in childhood and adolescence, slowing in adulthood so that new cells merely replace worn out or damaged ones. Puberty is a landmark along this journey, bringing about the changes necessary for adults to reach reproductive maturity.

20.1 Pregnancy and health

A medical leaflet given to a young woman said 'all women considering pregnancy should take supplements containing up to 4000 mcg of a B vitamin called folic acid, and ensure their diet includes foods which are naturally high in or fortified with folic acid'.

(a) Why is folic acid important for fetal development in early pregnancy? **[2]**

(b) What does mcg stand for? **[1]**

(c) Name a food naturally high in folic acid. **[1]**

(d) Name a food which is commonly fortified with vitamins. **[1]**

20.2 Growth

(a) What is growth? **[1]**

(b) Figure 20.1 shows two graphs using growth data. Explain what difference there is in the way that the data has been presented. **[2]**

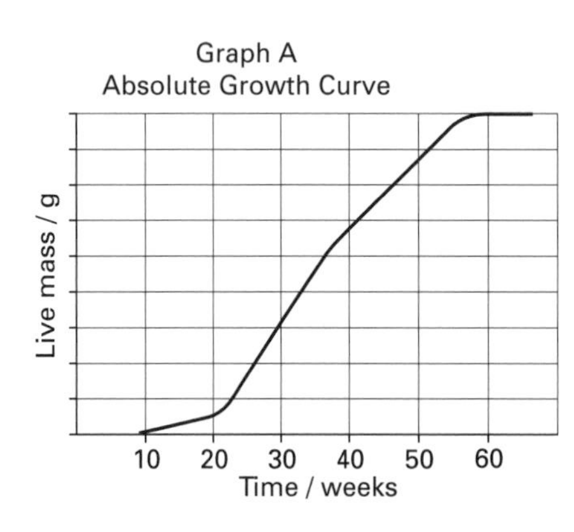

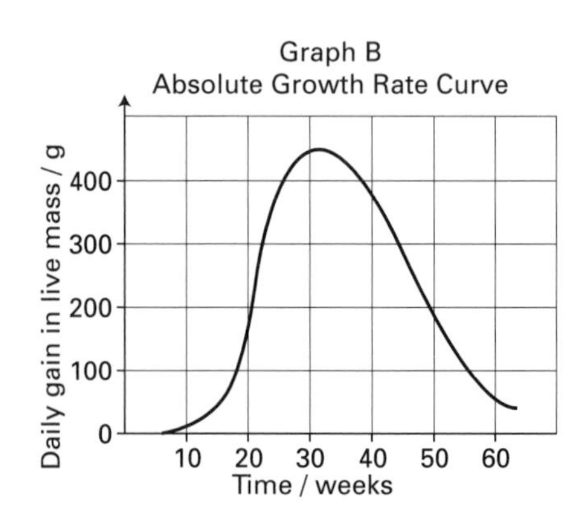

Figure 20.1

(c) How does growth occur at cellular level? **[3]**

(d) Name one tissue which grows throughout life and one which grows mostly during childhood. **[2]**

(e) Describe three ways in which the growth of an infant is measured. **[3]**

(f)

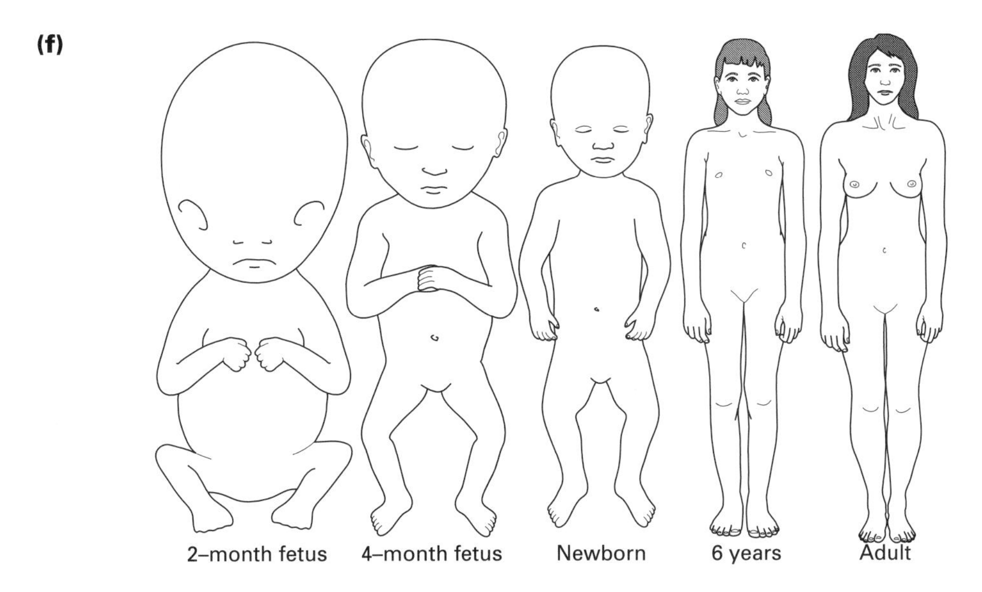

Figure 20.2 Comparative sizes of different parts of the body

Comment on the changing proportions of the body during life, as illustrated in Figure 20.2. **[4]**

20.3 Control of growth

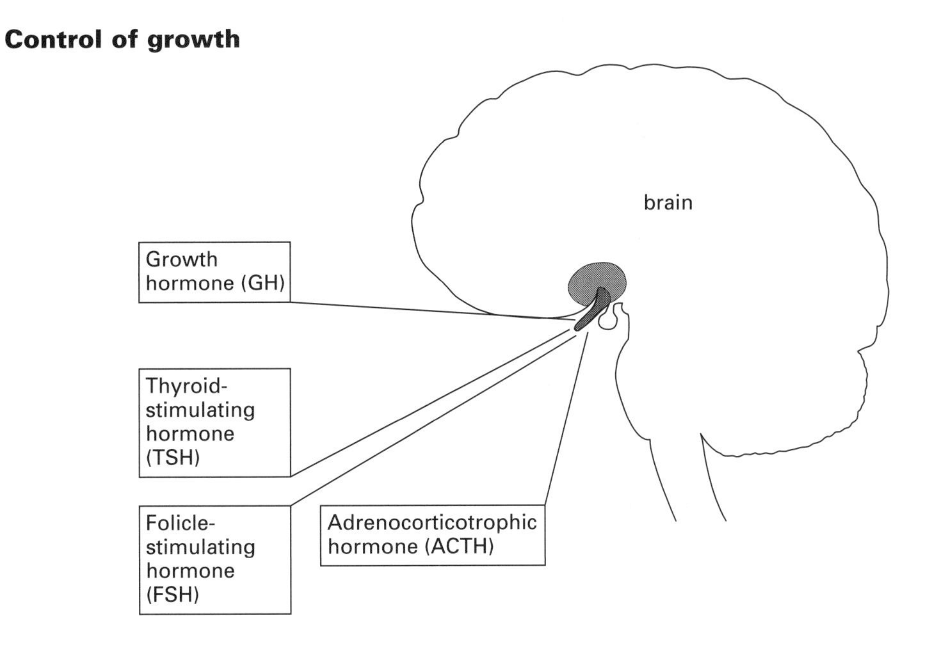

Figure 20.3 Hormones and growth

(a) Which two areas of the brain are involved in controlling growth via hormone release? **[2]**

(b) GH has the effect of increasing protein production. Explain why this enables and promotes growth generally. **[2]**

(c) Growth is retarded if insufficient GH is produced by the body. How is genetic engineering involved in solving this problem medically? **[2]**

(d) Which gland is affected by TSH, and how is it affected? **[2]**

(e) FSH does not affect growth of the whole body. Which tissue is affected by FSH, and at which stage in life is the major influence exerted? **[2]**

(f) How does ACTH help to make resources available for cell growth? **[2]**

20.4 Figure 20.4 shows the growth of a female from the age of 1 to 17 years.

(a) How can you tell from the graph when the most rapid phases of growth are? **[1]**

(b) When is the most rapid growth phase? **[1]**

(c) Comment on the growth rate at age 11–13. **[2]**

(d) Sketch on the growth curve for males. **[3]**

(e) Comment on the difference in growth between girls and boys aged 5–10 and 14–16. **[4]**

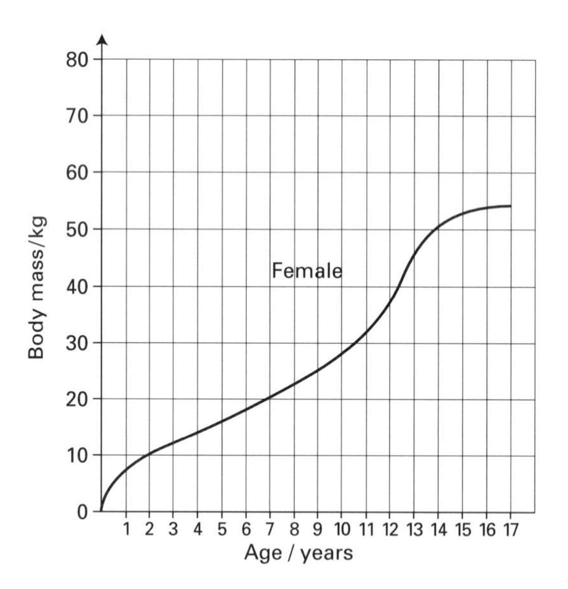

Figure 20.4 Growth curve for females: body mass

20.5 Puberty

Copy and complete the following passage by using appropriate words/phrases to fill the blank spaces.

Puberty happens during __________ as the body matures from childhood to adulthood. Various hormones bring about the changes, when their levels reach a __________. The __________ reproductive organs grow and the production of __________ begins. There is increased secretion of __________ including __________ in males and oestrogen in __________. The secondary sexual __________ appear as a change in body shape such as __________ in females, and changes in body tissues such as increased __________ development in males. Both sexes develop pubic hair and sweat glands become more active. **[10]**

20.6 Menstrual cycle

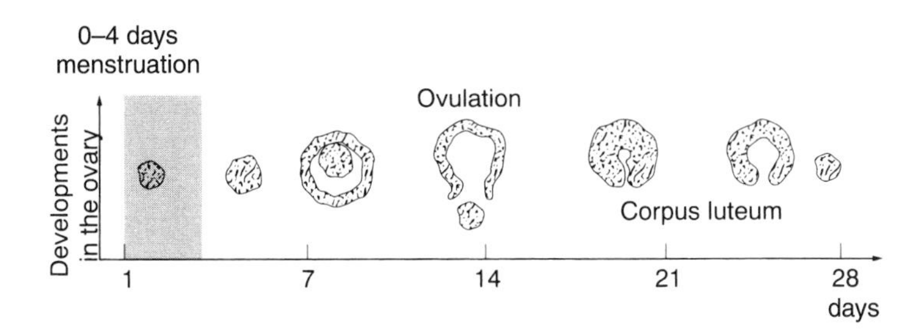

Figure 20.5 Events in the ovary during the menstrual cycle

(a) Which hormone influences the maturing of follicles in the ovary? **[1]**

(b) Mature follicles secrete oestrogen which has a negative feedback effect on the hormone that initially caused the follicle to mature. What is meant by a negative feedback effect? **[1]**

(c) Suggest a reason for the negative feedback effect referred to in **(b)**. **[1]**

(d) What is the function of the corpus luteum in the last part of the menstrual cycle? **[1]**

20.7

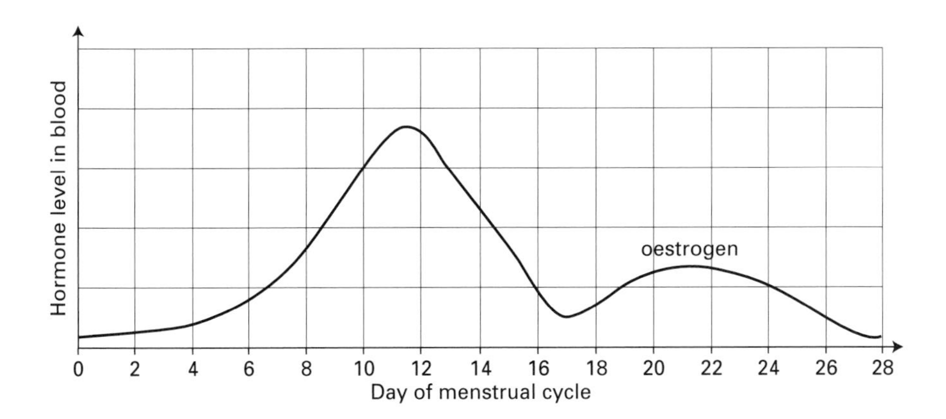

Figure 20.6 Concentration of oestrogen in blood

(a) Sketch a line on the graph to show progesterone level in blood during the menstrual cycle. **[2]**

(b) Which structure in the ovary produces progesterone? **[1]**

(c) Why does the production of progesterone decline towards the end of the cycle? **[2]**

(d) Under what circumstances does the level of progesterone remain high? **[1]**

20.8 Ageing

Write a short account of the general effects of ageing, including those on the skeletal system, the cardiovascular system and the reproductive system. **[10]**

20.9

(a) Distinguish between osteoporosis and osteoarthritis. **[4]**

(b) Name an acute neurodegenerative disease and a chronic neurodegenerative disease which may happen in old age. **[2]**

20.10 Mortality and longevity

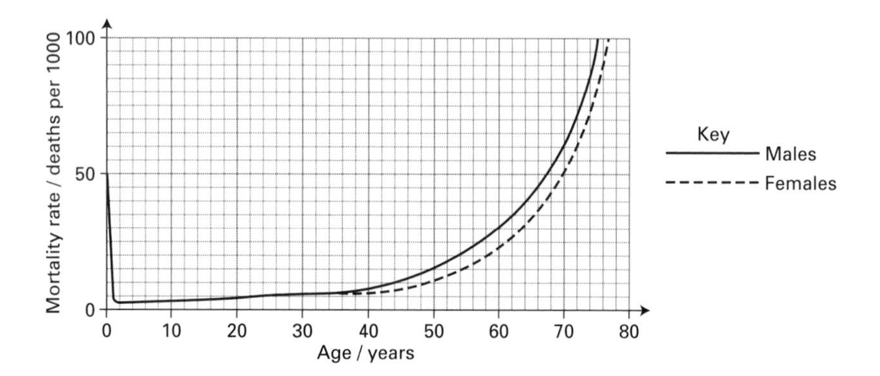

Figure 20.7 Mortality rate 1940 North America

(a) What is the mortality rate in 1940 for males aged 50? **[1]**

(b) Suggest a reason why the mortality rate for males is higher than for females of an equivalent age. **[2]**

(c) Describe how the mortality rate changes during childhood. **[4]**

(d) Suggest reasons for the infant mortality rate compared to older children. **[2]**

(e) Comment on the mortality rate for people between aged between 40 and 60. **[2]**

(f) How might this graph differ for North America in the year 2000? **[2]**

(g) The table shows data collated by the World Health Organisation (WHO) about how long people in different countries might expect to live without becoming seriously ill or disabled.

Life expectancy (adjusted for disability)

Ranking	Country	Years
1	Japan	74.5
2	Australia	73.2
3	France	73.1
4	Sweden	73.0
5	Spain	72.8
14	Britain	71.7
24	US	70.0
92	Russia	61.3
127	Iraq	55.3
184	Zimbabwe	32.9
191	Sierra Leone	25.9

(i) Suggest reasons why data adjusted in this way might be a more useful way of judging the health of a nation than life expectancy alone. **[2]**

(ii) Suggest two factors which might influence life span in countries such as Zimbabwe, which appear at the bottom of the table. **[2]**

(iii) The US is lower down the lifespan table than most other developed countries. Suggest what factors might most influence its position. **[2]**

21 Digestion and nutrition

Humans rely directly or indirectly on plants for all their food. A balanced diet providing carbohydrates, lipids, proteins, minerals, vitamins and water is essential to good health. Dietary fibre is also important. The exact requirement for each class of food varies with the age, sex, activity and state of health of the individual.

Proteins, carbohydrates and lipids are digested by enzymes to produce smaller soluble molecules that can be absorbed into the blood circulation for transport to cells in all parts of the body.

There are known health risks associated with both under nutrition (which may lead to anorexia nervosa) and over nutrition, causing obesity.

21.1 **(a)** Explain the terms: autotroph, heterotroph, and holozoic nutrition. **[3]**

(b) Distinguish between ingestion, digestion, and egestion. **[3]**

(c) Copy and complete the table which relates to some enzymes involved in digestion:

Site	Active enzyme	Substrate	Product(s)	pH
Mouth	salivary amylase			
Stomach	1 pepsin 2	 caseinogen		
Duodenum	1 2 3 pancreatic lipase		peptides maltose	 7.0
Ileum	1 2 3 maltase	lactose	 glucose fructose	

[22]

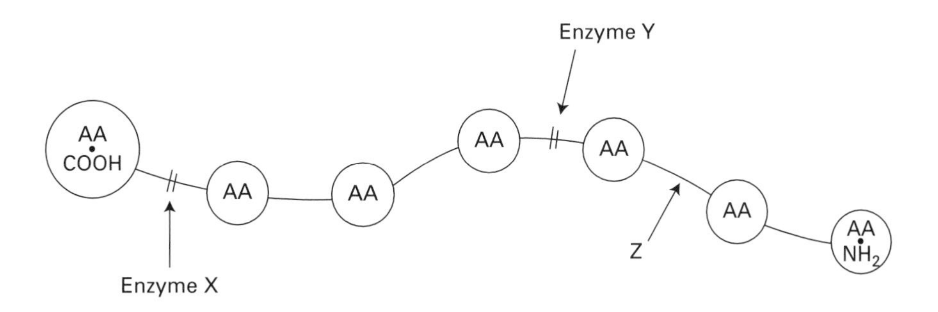

Figure 21.1

(d) (i) Identify X Y and Z in Figure 21.1. **[3]**

(ii) What are endopeptidase and exopeptidase enzymes? How do they digest polypeptides? Name an example of an endopeptidase. **[3]**

(e) Distinguish between extracellular and intracellular digestion in the ileum. **[2]**

21.2

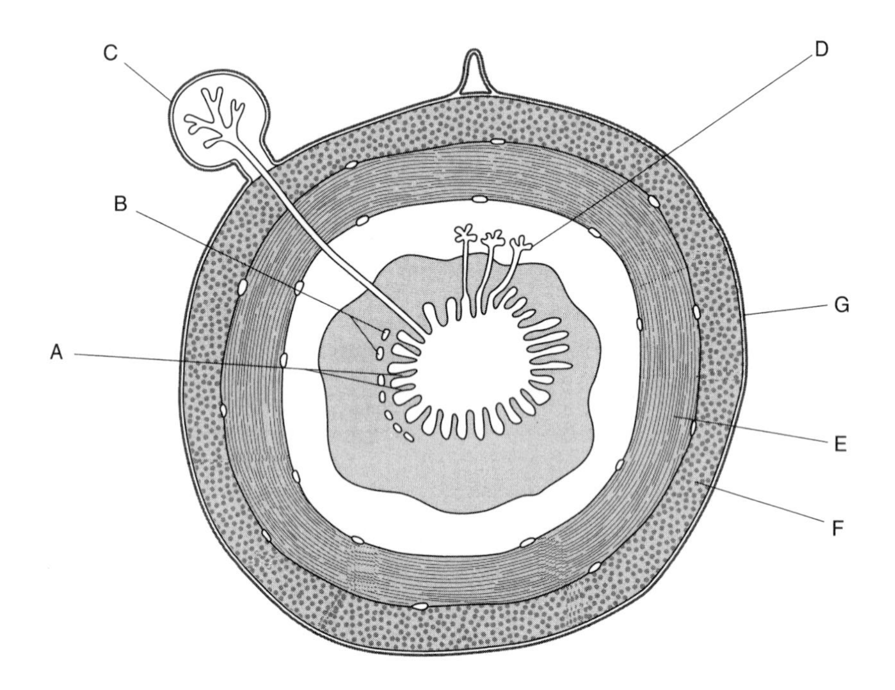

Figure 21.2

(a) What is A? **[1]**

(b) What is the function of parts B C and D? **[1]**

(c) Explain how parts E and F aid digestion. **[3]**

(d) What is G? **[1]**

21.3 **(a)** What part of the gut does Figure 21.3 show? **[1]**

(b) How is A structurally adapted for the efficient absorption of digested food? **[3]**

(c) How is a concentration gradient maintained across the inside and outside of A? **[3]**

(d) Explain how the following are absorbed by A and reach the blood system:

(i) glucose and amino acids **[3]**

(ii) fatty acids, glycerol and fat soluble vitamins **[3]**

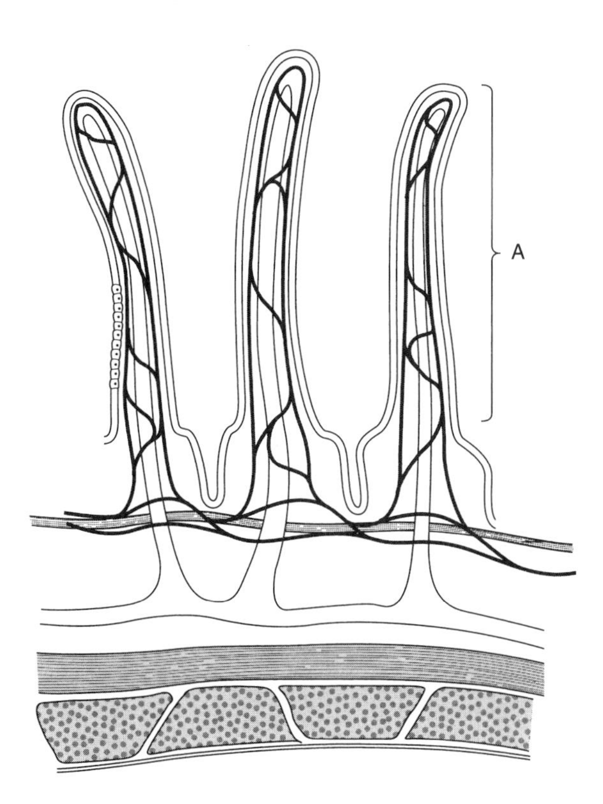

Figure 21.3

21.4 *Diarrhoea ranges from an inconvenient symptom to a debilitating or fatal condition. It has many causes all resulting in the loss of fluid and nutrients from the body.*

A healthy person will take in on average 1.5 litres of fluid daily, of which about 150 ml is expelled in the faeces. Large volumes of water are absorbed from the gut contents in the upper intestine, by osmosis. As nutrients and solutes from digestion are transported across the intestinal epithelium, the osmotic gradients necessary for water absorption are maintained. In the lower ileum and colon the active absorption of sodium chloride is responsible for the continued dehydration of the intestinal contents.

In the 1960s acute diarrhoea killed around 5 million children every year. The only effective treatment was to rehydrate the body via an intravenous drip. Giving the child extra water to drink did not help. It was then discovered that if glucose and salt were added to the water in similar proportions to those found in blood, liquid could be absorbed through the ileum. This treatment, Oral Rehydration Therapy has improved the situation significantly worldwide.

(a) Lactose-intolerant individuals are unable to secrete lactase, the enzyme that digests lactose. Why is diarrhoea a symptom of this condition? **[2]**

(b) Coeliac disease is characterised by abnormal villi in the small intestines. They are small and reduced in size. How does a structural defect like this cause diarrhoea? **[2]**

(c) Which sector of a population other than children is at risk if they develop severe diarrhoea? **[1]**

(d) Why was giving water to a sick child ineffective? **[1]**

(e) Give two disadvantages of using an intravenous drip to treat diarrhoea. **[2]**

(f) Suggest two advantages of Oral Rehydration Therapy. **[2]**

(g) Suggest one way in which the effect of the rehydration solution might be improved. **[2]**

(h) Explain why increasing the concentration of glucose and/or salt in the rehydration solution could be dangerous. **[2]**

21.5 Use words from the list to complete the following sentences.

secretin; enzyme; saliva; CCKPZ; reflex action; secretin; hydrochloric acid; hormone; cholecystokinin; enterogastrone; pepsinogen; trypsin; conditioned reflex

(a) The secretion of saliva is a ____________ the stimulus being the sight, smell and taste of food.

(b) Gastrin is a ____________ controlling the secretion of ________________ and ____________________ .

(c) Acid chyme entering the duodenum stimulates the secretion of ________________ and ____________________ .

(d) Liver cells secrete ____________ in response to the hormone ________________.

(e) ____________________ is the stimulus for the release of bile into the bile duct.

(f) In the pancreas ______________ stimulates the secretion of alkaline pancreatic juice and ______________ stimulates enzyme secretion.

(g) The secretion of acid in the stomach is inhibited by ____________. **[12]**

21.6 **(a)** Write a short paragraph to explain the term 'balanced diet'. **[5]**

(b) Name two biochemical constituents of dietary fibre. **[2]**

(c) List three functions of dietary fibre in the diet. **[3]**

(d) Give two reasons why people may follow a vegetarian diet. **[2]**

(e) Why may vegans be at risk from deficiency of vitamin B_{12} and calcium? **[2]**

Estimated Average Requirement (EAR) is a dietary reference value published by the UK Government. The table below shows the EAR for energy intake in MJ per day.

Age	Males	Females
0–3 months	2.28	2.16
Age (years)		
1–3	5.15	4.86
4–6	7.16	6.46
7–10	8.24	7.28
11–14	9.27	7.27
15–18	11.51	8.83
19–50	10.60	8.10
Over 75	8.77	7.61

(f) Why is the recommendation for the over 75s lower than for the 19–50 age group? **[2]**

(g) Explain the differences in recommendations for males and females. **[2]**

(h) Give three reasons why a government should publish dietary reference values such as EAR. **[3]**

21.7 **(a)** Define the term **basal metabolic rate (BMR)**. **[2]**

(b) Name two hormones that help to control BMR. **[2]**

(c) Copy the following table and tick (✓) true or false for each statement.

	True	False
Women have a faster resting BMR than men		
BMR is lower in older people		
BMR is higher in people living in a cold climate than in a temperate climate		
Following a serious injury BMR generally falls		
Metabolic rate rises after eating a meal		

[5]

(d) Account for the following:

(i) BMR increases during pregnancy. **[2]**

(ii) Regular exercise increases BMR. **[1]**

(iii) Average BMR in a group of moderately obese women was found to be higher than average BMR in a control group of lean women. **[2]**

21.8 Complete the blanks in the following passage.

All women planning a pregnancy are advised to take a daily supplement of __________ since it is difficult to achieve the extra recommended level by diet alone. Natural sources include __________ and fortified foods such as __________ .

During the first __________ months of pregnancy extra food is not usually necessary. However care is needed that __________ are restricted while ______________ are increased. These are needed for growth of the fetus and the __________ and for the increased volume of __________ in the body.

The __________ of nutrients is more efficient during pregnancy. Iron from red meat and ________________ need only be supplemented should the woman become ________________ . __________ has stopped and this helps to maintain levels of iron. Vitamin _______ aids the absorption of iron and so plenty of ______________ for example should be included in the diet. Calcium is another essential mineral, but again, a balanced diet should provide the recommended __________ mg each day. The metabolism of calcium is closely linked with Vitamin _____ obtained from food, and the action of the ____________ on the skin.

For the last three months of pregnancy the daily energy value of food intake should increase by around _________ (200 kcal) a day. However once the mother is breast feeding she will require extra _________ and carbohydrate to increase _________ and also calcium and __________ . About __________ ml milk are produced each day to feed the infant. **[22]**

21.9

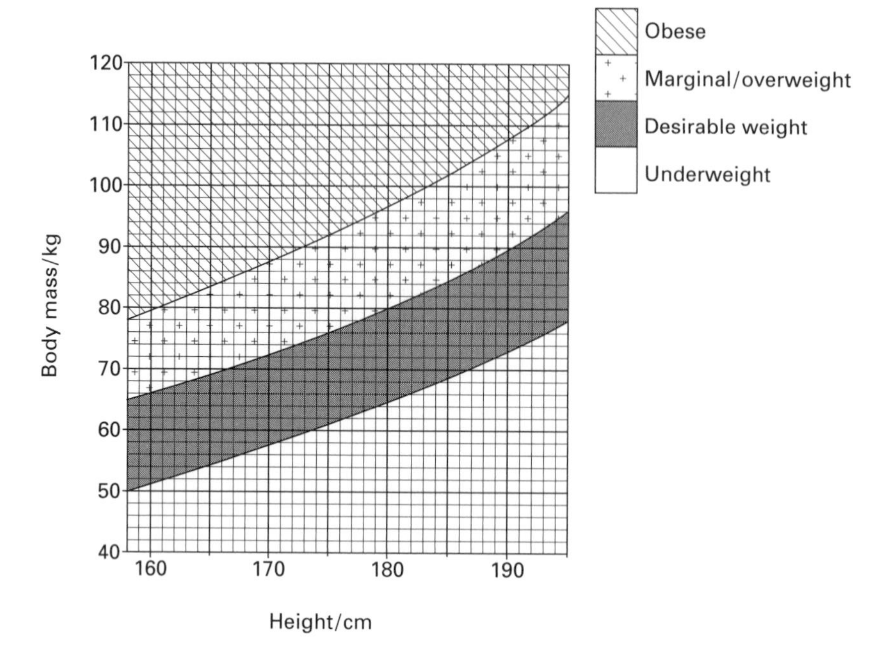

Figure 21.4

(a) A man weighing 80 kg is 1.78 m tall. Use the chart in Figure 21.4 to determine which category describes his body mass. **[1]**

(b) This man's body mass increases by 21% over a period of several years.

(i) Calculate his new body mass. Show your working. **[3]**

(ii) Body Mass Index (BMI) $= \dfrac{\text{mass/kg}}{(\text{height/m})^2}$

Calculate the BMI of the individual after he increased his body mass. Show your working. **[3]**

(c) What range of BMI is considered to be desirable? **[1]**

(d) Why is BMI a useful measure of obesity? **[1]**

(e) List four possible harmful ill effects of a high BMI. **[4]**

(f) **(i)** If an individual increased their daily exercise but not energy intake, why might their body mass not decrease? **[1]**

(ii) List three possible changes to the composition of their body tissues. **[3]**

(g) What eating disorder might an individual with a BMI of 15.5 be suffering from? **[1]**

21.10 Write a short account of the symptoms and causes of *anorexia nervosa*. How does this disorder differ from bulimia? **[10]**

22 Homeostasis: principles and examples

Cells and body systems function best under certain conditions, such as temperature, oxygen supply, glucose levels, pH and salt balance. Despite fluctuations in terms of inputs and outputs of these conditions and many others, the body is able to control these conditions within a fairly narrow range.

The liver is responsible for controlling glucose level in blood, the kidneys balance water and salt level, and the skin is mainly responsible for controlling body temperature.

In response to changes brought about by varying levels of activity, the breathing rate and heart rate can be adjusted.

22.1 **(a)** What is meant by the terms:

(i) external cell environment ? **[1]**

(ii) internal cell environment? **[1]**

(b) Name two conditions which can vary in the external cell environment. **[2]**
Why is each condition important for cell functioning? **[2]**

(c) pH is an internal cell condition. Why is the correct pH essential for a cell to function properly? **[2]**

22.2 **Homeostasis**

Figure 22.1 is a model of the concept of homeostasis. The line represents a condition within the human, such as body temperature.

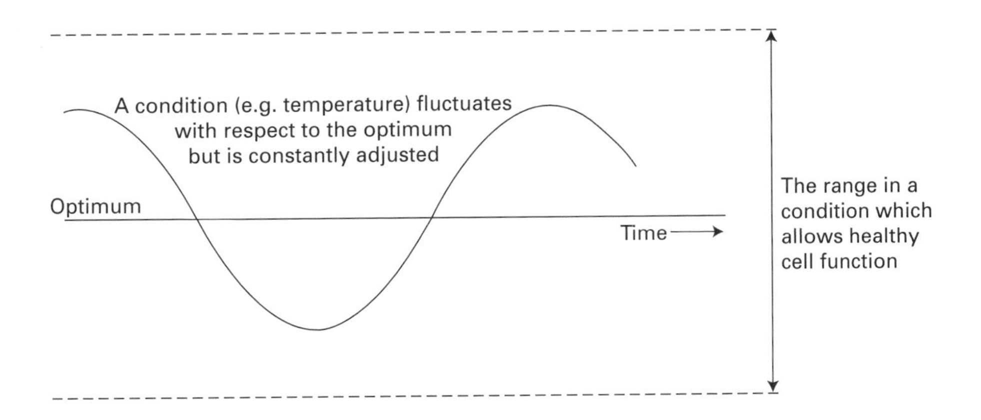

Figure 22.1

(a) Explain why the line fluctuates away from the optimum and back towards it. Include the term **negative feedback** in your answer. **[4]**

(b) How do cells detect changes such as a drop or rise in the level of glucose in blood? **[1]**

(c) What is meant by an effector organ? **[1]**

22.3 Copy and complete the following table by placing a tick in the spaces for true statements and a cross for false statements.

	Main organ systems involved in homeostasis					
Statements	**Kidney**	**Hypothalamus**	**Skin**	**Liver**	**Pituitary**	**Thyroid**
Acts as a master gland						
Regulates water, salt and pH in body fluids						
Brings about changes in body temperature						
Controls the pituitary gland						
Regulates the metabolic rate						
Regulates carbo-hydrate metabolism						

[6]

22.4 **Blood glucose level**

Define the terms hyperglycaemia and hypoglycaemia. **[4]**

22.5 Compile a table of information about the effects of insulin and glucagon. Include their effects on the rate of cellular respiration, as well as carbohydrate and lipid interconversion. **[8]**

22.6 Write an account of the functions of the liver. **[10]**

22.7 **(a)** What is the likely long-term effect on the liver of regularly drinking alcohol to excess? **[3]**

(b) Name two toxins other than alcohol which are broken down by the liver. **[2]**

(c) The most rapid phase of liver growth is in the fetus and in infancy. Under what other conditions may liver growth be rapid? **[2]**

(d) One observable effect of liver dysfunction is jaundice. Explain what this is and why it occurs. **[2]**

(e) Describe the main blood supply to and from the liver. **[2]**

22.8 Osmoregulation

Copy Figure 22.2 which shows an outline of the kidney structure. Name the parts indicated and write a sentence about (annotate) each. **[20]**

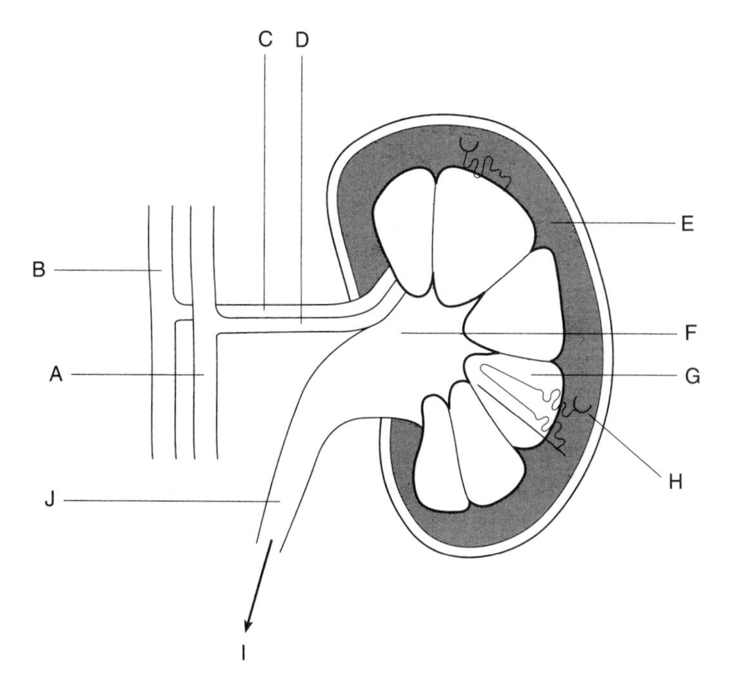

Figure 22.2

22.9

Complete the passage by filling the blanks with appropriate words or phrases.

The kidneys have the role of balancing water and _________ content in blood, as well as the _________ of urea and helping to maintain the _________ of blood. The initial stage of all these processes happens in the Bowman's capsule, where _________ occurs. The blood pressure forces substances such as urea, water, ions, vitamins, hormones and _________ into the tubule. Some molecules and blood cells are held back because of their larger _________.

_________ happens in the first convolluted tubule, causing over 80% of the filtered materials to pass back into the blood. This is an active process, and the epithelial cells of the tubule are packed with _________ as they transfer the required energy.

The role of the loop of Henle is to concentrate the filtrate. This happens when water passes out by _________ along a concentration gradient, into the interstitial tissues of the medulla and the capillaries of the _________. Sodium and chloride ions are pumped out of the ascending limb by _________, which lowers the _________ of the medulla and maintains the concentration gradient.

Finally, the fine adjustment of ion concentration occurs in the _________ and collecting ducts. The hormone _________ is important in this process. **[14]**

22.10

	Volume of water reabsorbed in different parts of the nephron (dm^3/day)
First convoluted tubules	147.5
Loops of Henle	10.0
Second convoluted tubules	19.0
Collecting ducts	2.5

The table shows the results of an investigation recording the volumes of water reabsorbed from the filtrate, on a particular day when the total volume filtered was 180 dm^3.

(a) Calculate the volume of urine produced that day. **[2]**

(b) The following day the person being investigated had the same volume to drink but was required to exercise regularly throughout the investigation. Suggest how the volume of urine might differ, and why. **[2]**

(c) Explain why eating a meal with a high salt content makes you feel thirsty. **[3]**

22.11 Figure 22.3 shows a dialysis machine which can be used by someone with kidney failure.

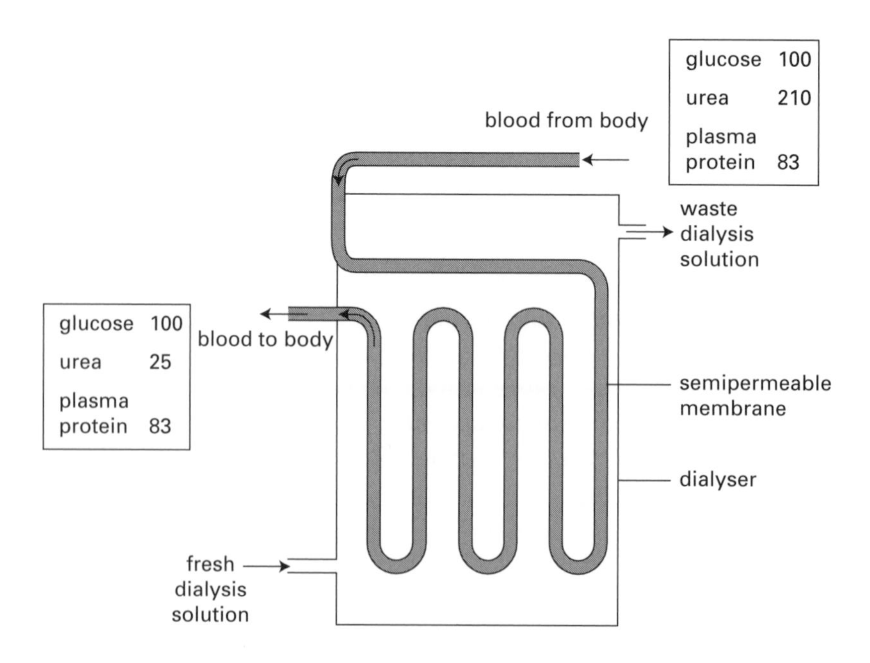

Figure 22.3 A dialysis machine

(a) Explain why the level of urea changes between the blood leaving the body and being returned to it. **[2]**

(b) Why does the level of glucose remain the same? **[2]**

(c) Why does the level of plasma proteins remain the same? **[2]**

(d) Comment on the fact that the dialysis solution and blood flow are passing in opposite directions. **[2]**

(e) The blood passing back into the body passes through a trap before returning. Explain one reason for the trap being vital in this circuit. **[2]**

22.12 **(a)** What are the main ways in which heat is lost and gained by the body? **[5]**

(b) What is meant by core temperature? **[1]**

(c) Under normal circumstances, what is human core body temperature? **[1]**

(d) Suggest two reasons why core temperature might rise above the normal level. **[2]**

(e) Why do the body extremities cool before the body core? **[2]**

(f) Body temperature exhibits diurnal fluctuation. What does this mean? **[1]**

(g) When would you expect body temperature to be at its lowest, and why? **[3]**

22.13 **(a)** Write a sentence about each of these features of skin: melanin; photoageing; melanoma; scar tissue; erector pili muscle; sebaceous gland. **[12]**

(b) Figure 22.4 shows the capillary circulation in the skin in cold conditions.

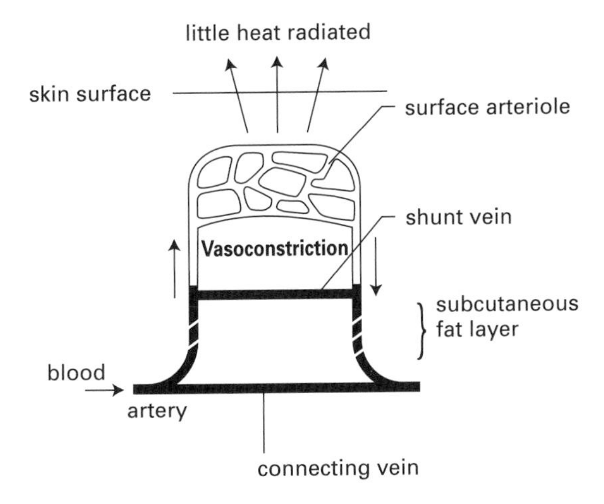

Figure 22.4 Blood supply to the skin

Sketch a similar diagram to show what happens to the capillary circulation under hot conditions. **[3]**

(c) Describe hypothermia and give examples of vulnerable groups of people. **[5]**

(d) Explain how the heat gain centre and heat loss centre of the brain are involved in thermoregulation. **[6]**

23 Nervous system

The nervous system is the major body system involved in detecting stimuli, coordination and response. Receptor cells detect changes within and outside the body, and pass information in the form of impulses via the central nervous system (CNS) to effector organs. The CNS acts as a processing centre, consisting of the brain and spinal cord.

Both muscles and glands are effector organs that respond by acting on the information they receive.

23.1 Neurones

(a) Figure 23.1 shows two simplified diagrams of nerve cells. Copy the diagrams and label and annotate features A–E. **[10]**

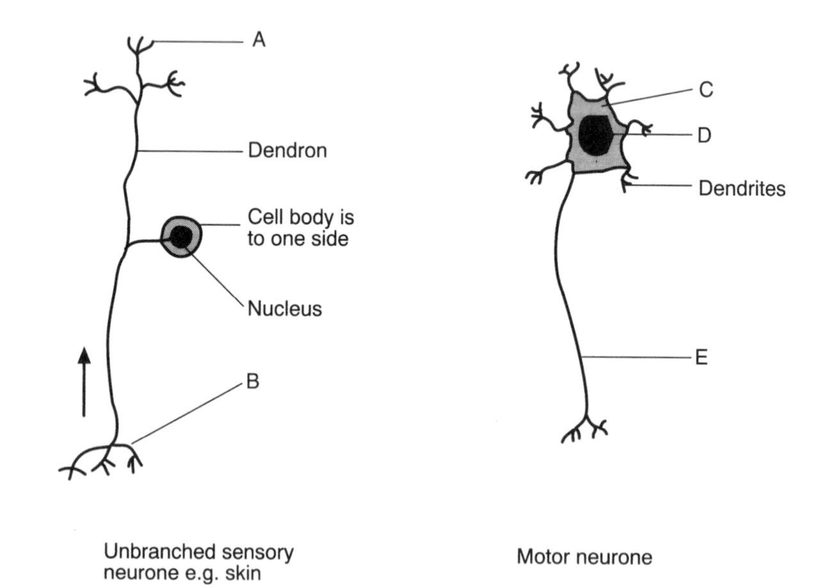

Figure 23.1

(b) Mark on your drawings the direction of impulse transmission. **[2]**

(c) Define these terms: **neurone, stimulus, receptor, neuroglial cell, Schwann cell.** **[5]**

(d) Explain the difference between a primary and a secondary receptor. **[2]**

23.2

Write a brief account of the general structure of the nervous system. You need to include these terms: **receptor/sensory cell, peripheral nervous system, central nervous system, motor cell, effector organ.** **[10]**

23.3 Nerve impulse transmission

Figure 23.2 shows the potential difference across the membrane of a neurone in between impulses.

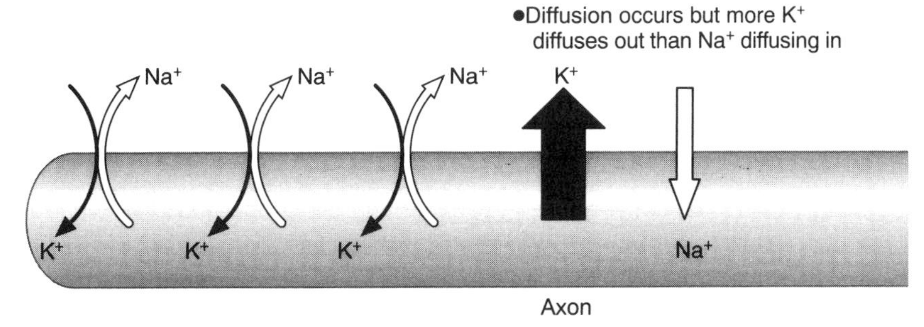

Figure 23.2 A resting neurone

(a) How do sodium and potassium ions move across the membrane? **[1]**

(b) Why do more potassium ions diffuse out of the neurone than sodium ions diffuse in? **[1]**

(c) What is the polarity of the membrane when the neurone is not transmitting an impulse? **[1]**

(d) How do the sodium gates alter the polarity as the impulse passes along the neurone? **[2]**

23.4

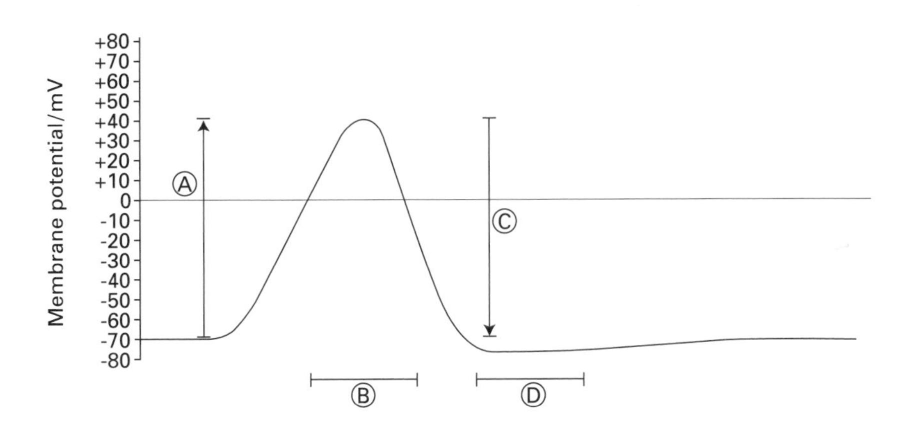

Figure 23.3 Transmission of an impulse

Match terms from the list below to each of the areas of the diagram marked A–D.

Select words from this list: **polarisation, depolarisation, repolarisation, apolarisation, gestation period, refractory period, fraction potential, gustation period, action potential.** **[4]**

23.5

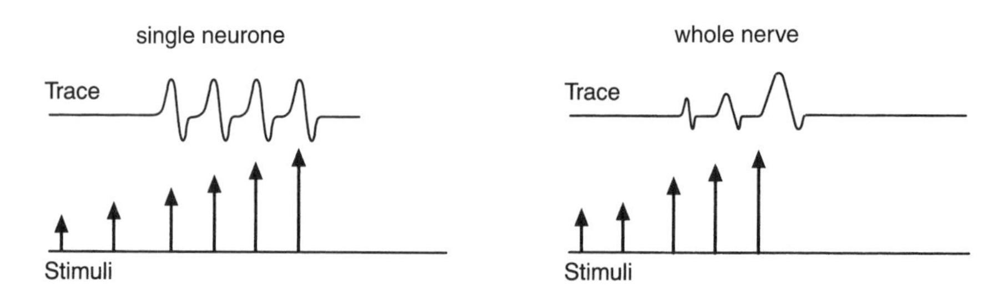

Figure 23.4 Stimulation of a single neurone and a whole nerve

Use Figure 23.4 to illustrate what is meant by:

(a) the threshold point **[1]**

(b) the all or nothing rule **[1]**

(c) how the response of a whole nerve differs from that of a single neurone. **[2]**

23.6 **(a)** What is a synapse? **[1]**

(b) What is a neurotransmitter? **[1]**
Give one example of a human neurotransmitter. **[1]**

(c) Explain the difference between an inhibitory and an excitatory neurotransmitter. **[2]**

(d) Describe accommodation (or adaptation) of receptors on sensory neurones. **[3]**

(e) Describe the effects of agonistic and antagonistic drugs on the synapse and name one example of each. **[4]**

23.7 Figure 23.5 shows a synapse between two neurones.

(a) Sketch a model of the interface between nerve cell and muscle. **[5]**

(b) Compare a synapse between two neurones and a neuromuscular junction. **[4]**

(c) Name one disease which affects neuromuscular junctions, preventing proper functioning. **[1]**

(d) What is meant by muscle tone and how can it be increased? **[2]**

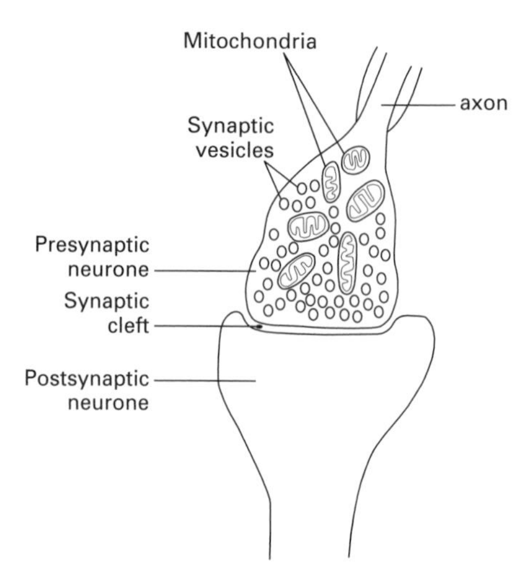

Figure 23.5 .

23.8 When an impulse arrives at a synapse there are changes to the membrane potential.

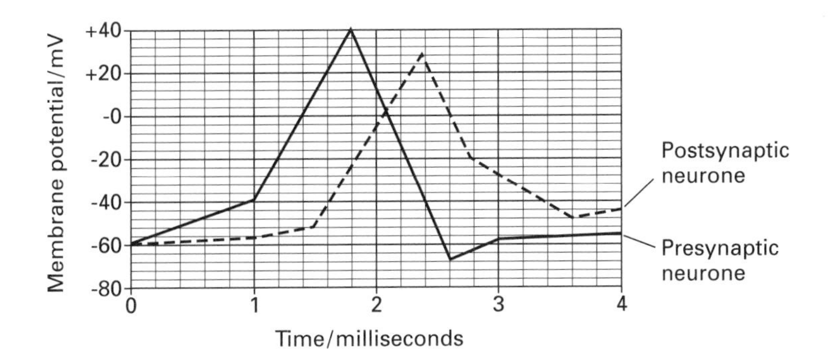

Figure 23.6 Changes in membrane potential

(a) Describe the change in membrane potential of the presynaptic neurone. **[5]**

(b) What causes the change in polarity at 1.8 ms? **[2]**

(c) What is the time delay between depolarisation of the presynaptic and postsynaptic membranes, and why does it occur? **[3]**

(d) How does caffeine affect synaptic transmission? **[2]**

23.9 Central nervous system

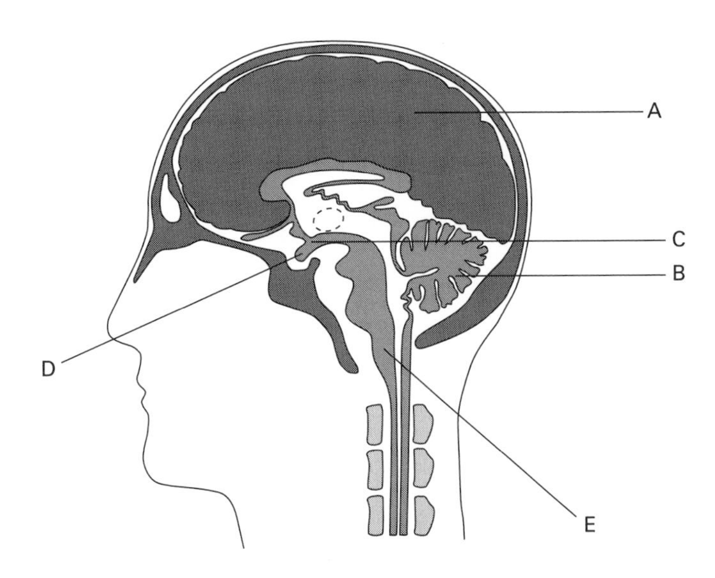

Figure 23.7 Cross-section of the brain

(a) Label the features A–E. **[5]**

(b) In what ways has knowledge of the functioning of the human brain been gathered? **[3]**

(c) Copy an outline of the brain and sketch on the areas concerned with: **motor control, memory, thought processes.** **[3]**

(d) Draw a flow chart showing the overall organisation of the nervous system. **[5]**

23.10 Copy and complete the table concerning the spinal cord and reflex actions by ticking true or false for each statement

Statement	True	False
Sensory neurones enter the spinal cord through the dorsal root		
A dorsal root ganglion is a collection of motor neurone cell bodies		
The white matter is mostly made up of collagen tissue		
The grey matter is composed of relay neurones and synapses		
Sensory nerves link the spinal cord to the brain		
An unconditioned reflex is modified by experience		
Swallowing is an unconditioned reflex.		

[7]

23.11 Effector organ

Write a brief account of the general structure of the eye, explaining how some of the structures are involved in focusing. **[10]**

23.12

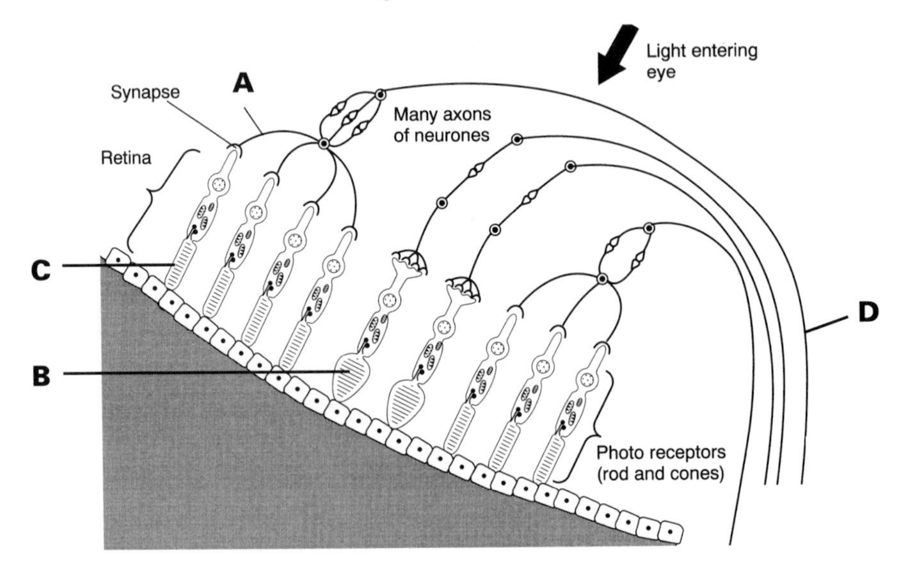

Figure 23.8 The retina

(a) Which of the structures is:

(i) a bipolar neurone? **[1]**

(ii) a rod? **[1]**

(iii) a cone? **[1]**

(iv) forms part of the optic nerve? **[1]**

(b) Explain why cones function best in high light intensities while rods can function in dim light. **[5]**

(c) What is colour blindness and why does it occur? **[2]**

23.13 Co-ordination

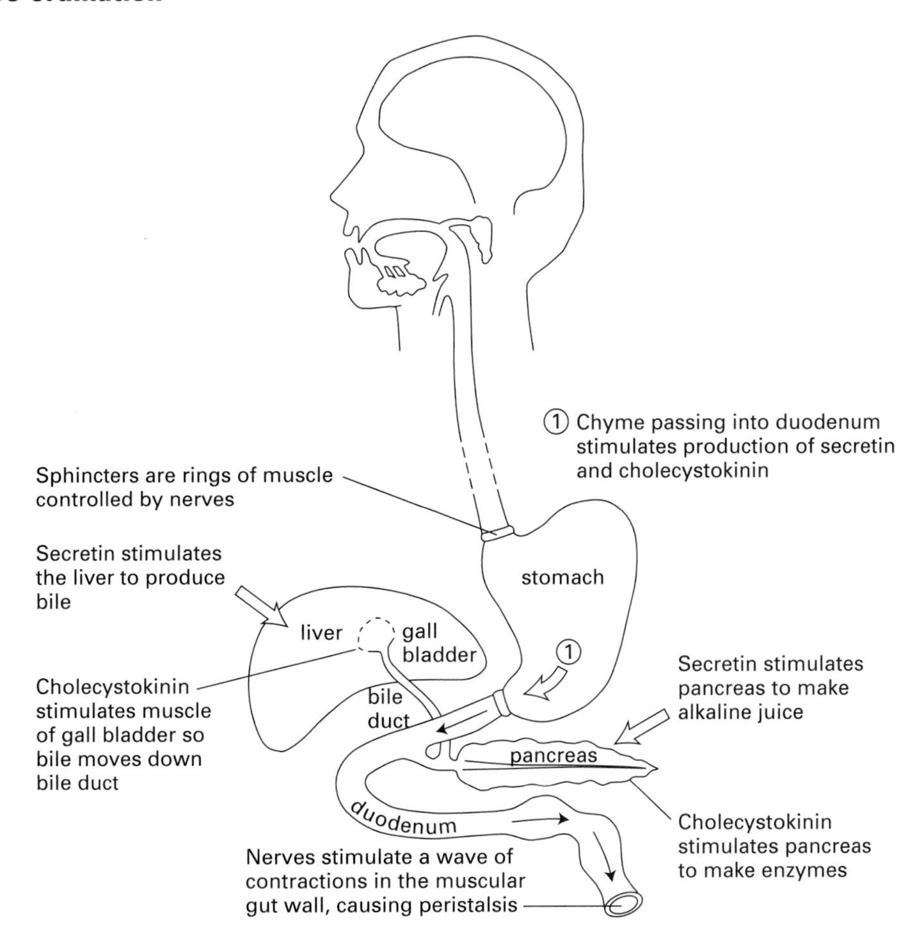

Figure 23.9 Controlling function of the alimentary canal

Use Figure 23.9 to explain how both the nervous and hormonal systems interact to bring about proper functioning of the gut. **[10]**

23.14 Drugs and the nervous system

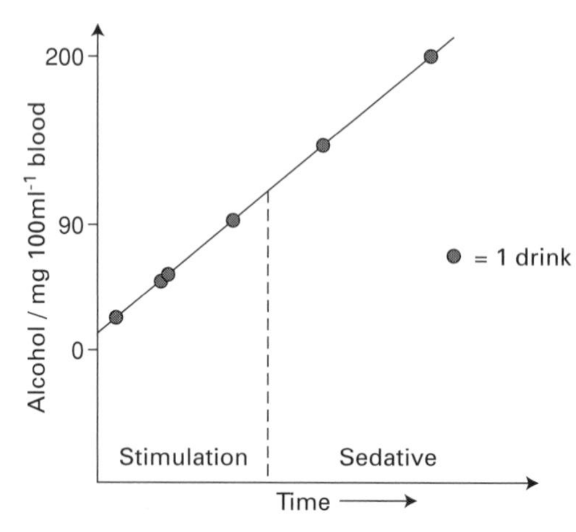

Figure 23.10 The relationship of blood alcohol level and effect on the nervous system

Figure 23.10 charts the progressive effects of an evening's consumption of alcohol for a partygoer. Around 25 mg/100 ml of blood sensitises one of the brain's major excitatory message pathways (called NMDA system), making certain receptors more sensitive to the neurotransmitter glutamate. This has the effect of raising brain activity, particularly in the areas associated with thinking, pleasure and reward, and motor control, which results in lowered inhibitions and increased confidence.

At around 80 mg/100 ml the partygoer's mood is heightened, and he/she is having a great time. With boosted alpha brain waves and extra blood flow to the prefrontal cortex and right side of the brain, the mood becomes euphoric. Now, the same receptors which were first stimulated by alcohol are not responding but, instead, another inhibitory pathway (called GABA system) is activated. This dulls brain activity by blocking impulse transmission and slowing the responses of neurones. At this stage alcohol acts as a sedative. The partygoer feels very relaxed and later, with even higher alcohol levels in the blood, experiences loss of coordination, loss of memory and finally sleep.

(a) Use Figure 23.10 to estimate the alcohol level in blood of someone who does not consume alcoholic drinks. **[2]**

(b) Why does consuming one alcoholic drink make this partygoer feel less nervous about meeting people at the party, and getting onto the dance floor? **[2]**

(c) Suggest reasons why exceeding the British legal limit for blood alcohol (80 mg/100 ml) is dangerous. **[2]**

(d) At what alcohol level does the NMDA system stop responding and the GABA system become activated? **[1]**

(e) Why can alcohol act as a sedative as well as a stimulant? **[4]**

(f) Scientists have found that the brain of someone who has a blood alcohol level of around 80 mg/100 ml uses about 25% less glucose than someone who has not had an alcoholic drink. How might this contribute to increased response time and reduced motor control? **[2]**

24 Effectors: muscles and glands

Muscles and glands bring about changes in the body when they are stimulated by impulses reaching them. There are different types of muscles which have different functions within the body and their structure varies accordingly.

Glands produce hormones that act on the target organs and tissues which have specific receptors.

24.1 Muscles

Effectors bring about change within the body when stimulated by impulses from a motor neurone. They include glands and muscles.

(a) Describe three changes which are brought about by muscles. **[3]**

(b) Describe three changes brought about by glands. **[3]**

(c) How is it possible that stimulating an organ such as the liver can bring about two different responses? **[1]**

24.2 Copy and complete the following table.

Feature	Muscle type		
	Smooth unstriated	**Striated**	(1)
How it is stimulated	autonomic nervous system	(2)	(3)
Microscopic appearance	(4)	elongated banded fibres	interconnected network
Activity	Slow to fatigue, involuntary	(5)	rhythmical contractions
Example of where it's found	(6)	(7)	(8)

[8]

24.3 Draw a simple diagram of the bones and main muscle pairs in the human arm, and explain how movement of the limb occurs. [10]

24.4 This is an extract from an article about body size in gymnasts.

This year, Nick Grantham of Lileshall Sports Injury and Human Performance Centre in Shropshire examined the body profiles of top British athletes… Men averaged 1.64 metres (5′ 5″) and women averaged 1.48 metres (4′ 8″)…. Being short has several advantages. Once you're in the air, if you're short, it's certainly much easier to somersault … once you've set up your rotation, the smaller you can become, the faster you can go – you can fit more somersaults in.....Long legs are long levers (and) you have to work harder to move them.

(a) How do limbs act as levers? Use a simple sketch to help your explanation. **[5]**

(b) What other advantages might a small gymnast have over a taller one, other than being able to fit in more manoeuvres? **[2]**

(c) What disadvantages might there be for a smaller gymnast? **[2]**

24.5 Briefly explain the function of each of the following skeletal tissues.

(a) bone **[2]**

(b) cartilage **[2]**

(c) ligament **[2]**

(d) tendon **[2]**

24.6 Copy the passage and fill in the blanks with appropriate words or phrases.

The sliding filament theory is a model to describe how muscle __________ is brought about. Each muscle fibre consists of myofilaments composed of two proteins: __________ and myosin, which overlap partially. _________ ions help to unblock binding sites on the thinner filaments so that they become linked with the myosin filaments, via myosin __________. This causes the filaments to slide in between each other, shortening the length of the muscle fibre overall. __________ generated during respiration transfers energy to the contracting fibres. This later reforms in the presence of __________ and phosphate ions. **[6]**

24.7 **(a)** What is the difference between a slow and a fast twitch muscle fibre in terms of speed of contraction, and the rate at which they fatigue? **[4]**

(b) What is myoglobin? **[1]**

(c) Which type of muscle fibre mentioned in **(a)** contains more myoglobin? **[1]**

(d) How does myoglobin contribute to muscle function? **[2]**

(e) How does regular physical exercise alter the proportion of these two types of fibre in a muscle? **[1]**

(f) What other skeletal difference might an athlete have to someone who avoids exercise? **[1]**

(g) Which other main body system benefits from regular exercise? **[1]**

(h) How might a sprinter use diet to improve performance? **[2]**

24.8 **(a)** What is a muscle spindle? **[3]**

(b) How does the information from muscle spindles reach the brain? **[1]**

(c) What is the result of stimulating muscle spindles? **[2]**

(d) Why are muscle spindles important in getting the right level of muscle action? **[2]**

(e) What is meant by muscle tone? **[1]**

(f) How does muscle tone help in maintaining posture? **[1]**

24.9 Muscles and exercise

Imagine that a student runs to the bus stop, starting at midday. Figure 24.1 records the change in oxygen consumption over that period of time, and for a while after.

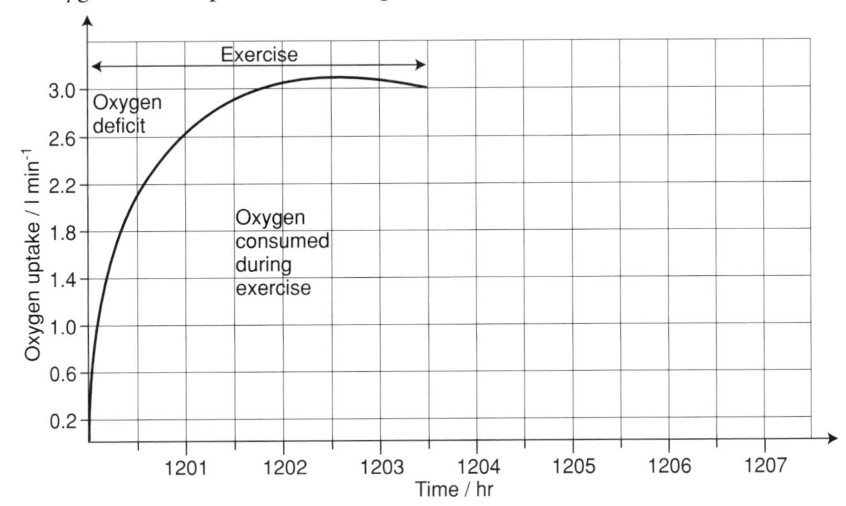

Figure 24.1 Oxygen uptake during exercise

(a) What is her level of oxygen consumption at midday? **[1]**

(b) Describe changes in the level of oxygen consumption between 1200 h and 1203 h. **[3]**

(c) What ventilation changes happen during this time? **[2]**

(d) How does the process of respiration alter during this time? **[1]**

(e) The student sits down once she catches the bus at 1205 h.Copy the diagram and sketch the shape of the graph you would expect over the next five minutes. **[2]**

(f) Why does breathing rate take a few minutes to return to the normal rate for resting? **[3]**

24.10

Here is an extract from an article about stress and osteoporosis.

... Bad diet or eating disorders are already known to weaken girls' bones. Now it seems that worrying about weight can release hormones that lead to weaker bones .....

Susan Barr and Jerilynn Prior (in) Canada quizzed 51 healthy pre-teen girls, none of whom had eating disorders, about weight worries. They also measured the mineral content of the girls' bones using a low dose of X-rays. To their surprise, they found that the girls who were more concerned about their weight than average were significantly more likely to have lower bone densities. ... (Their) work ...suggests that college girls who fret too much have higher than usual levels of urinary cortisol, a stress hormone which has a negative effect on bone development. While it's a bit of a leap to relate those results to pre-teens, says Barr, the same thing may be happening with weight worries .....

(a) Why have the authors of this article called cortisol a stress hormone? **[1]**

(b) Which of the survey techniques described in this article is quantitative, and which provides qualitative information? **[3]**

(c) Which type of data is most likely to be used in a scientific report as evidence to support an idea or theory? **[2]**

(d) Suggest reasons (other than anxiety) for lower bone densities in some females. **[3]**

(e) If you were organising this type of survey, what other variables might you need to consider? **[2]**

(f) What does Barr mean by 'a bit of a leap' in the last sentence? **[2]**

(g) Why is bone density an important health factor? **[3]**

(h) Suggest two ways of improving bone density. **[2]**

24.11 Endocrine system

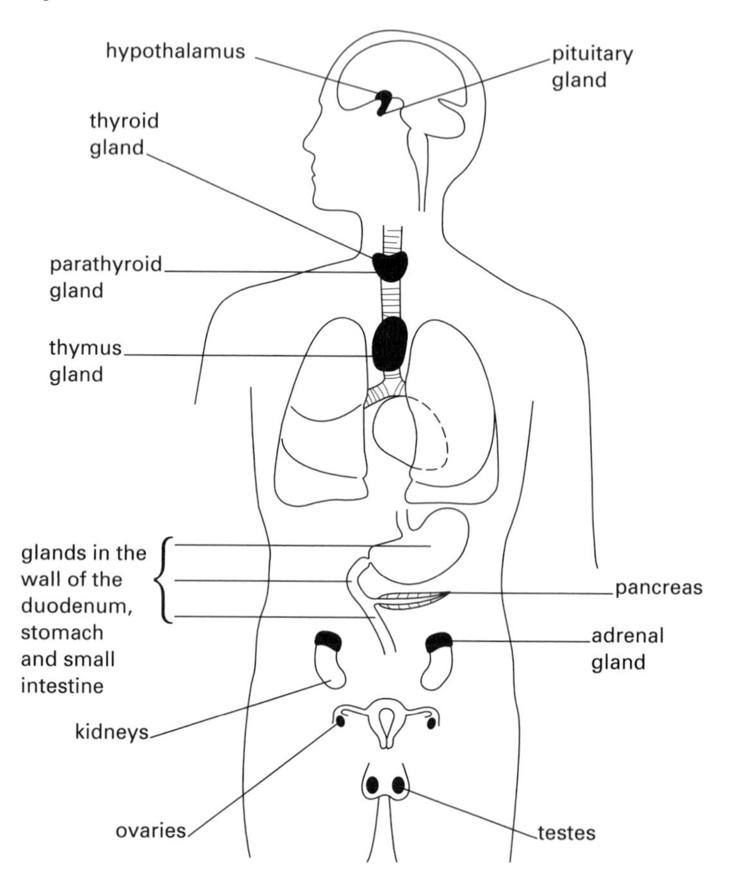

Figure 24.2 The main hormone producing organs and tissue

Use Figure 24.2 to help you to tabulate a summary of information about the main human hormones. You need to include the name of the hormone, the source, the target organ/tissue and the effects it causes. **[3 per hormone]**

24.12 Write a short account of the main ways in which hormones affect cells, giving examples where possible. **[10]**

24.13

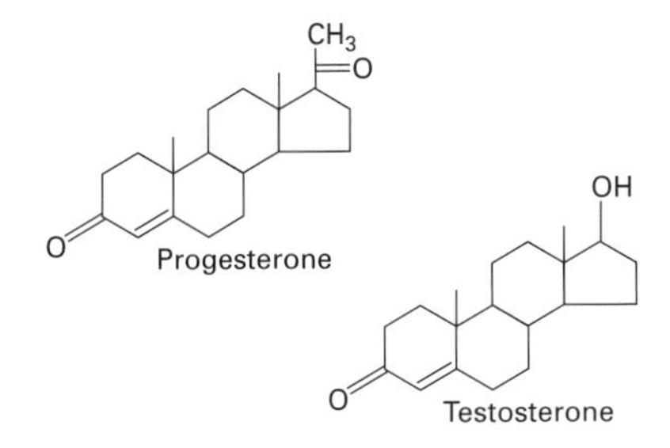

Figure 24.3 The structure of progesterone and testosterone

(a) Comment on the structure of these hormones. **[3]**

(b) What type of hormones are these? **[1]**

(c) Both these hormones are lipid soluble and cannot be stored by cells within vesicles. Why not? **[2]**

(d) Suggest why the starting point for many medicines is substances that are found naturally in living systems. **[2]**

24.14 One of the symptoms of diabetes is the production of large volumes of dilute urine. The graph shows the effect of pituitary extract on the volume of urine produced.

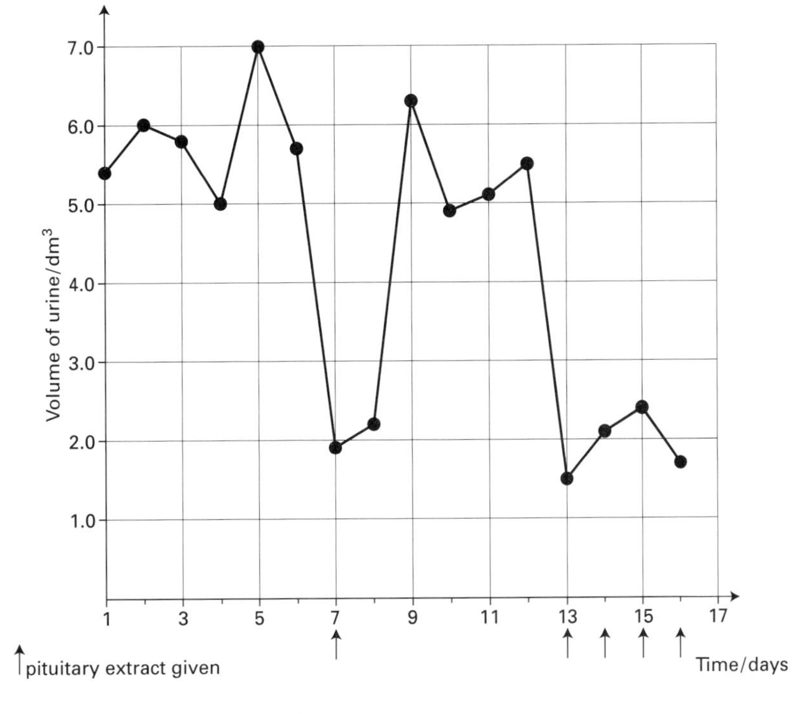

Figure 24.4 Volume of urine production

(a) What is the effect of pituitary extract on urine production? **[1]**

(b) What substance in the extract causes this effect? **[1]**

(c) How does this substance cause this effect? **[2]**

(d) What treatment is usually prescribed for diabetes? **[2]**

(e) What is gestational diabetes? **[1]**

Answers

1 Cells

1.1 **(a)** **A** flagellum; **B** nucleus/nucleoid/nuclear material; **C** cell wall; **D** cell membrane; **E** mesosome; **[5]**

(b) (muco)polysaccharides and amino acids **[1]**

(c) glycogen; lipid; **[2]**

(d) protection against host's immune system/ digestive enzymes; **[1]**

(e) colony; **[1]**

1.2 **(a)** bacterium/bacteria/bacterial cell; named plant/ plant cell; named animal cell/animal cell; **[2]**

(b)

Feature	Prokaryotic cell	Eukaryotic cell
Nucleus	✗	✓
Mitochondria	✗	✓
Mesosome	✓	✗
Ribosome	✓	✓
Cell surface membrane	✓	✓
Endoplasmic reticulum	✗	✓
Plasmid	✓	✗

[7]

Note: in this kind of question you must have the complete row correct, to gain each mark.

1.3 **(a)** **A**=cell membrane; **B**=mitochondrion; **C**=secretory vacuole; **D**=nuclear membrane; **E**=nucleoplasm/chromatin/nucleus; **[5]**

(b) $\text{Magnification} = \frac{\text{observed size}}{\text{actual size}}$

(This may help)

Measurement from drawing (O) = 7 mm

Magnification (M) = × 2000
(See first line of question)

Actual size (A) = 7 ÷ 2000 = 0.0035 mm

1 mm = 1000 μm

ANSWER Actual length of B is 3.5 μm **[2]**

(c) **(i)** controls exchange of molecules and ions between cell and environment; recognition/communication; **[1]**

(ii) matrix is site of Krebs Cycle/cristae are site of electron transfer chain; production of ATP by oxidative phosphorylation; site of fatty acid metabolism; **[1]**

(iii) production of glycoproteins/mucus; processing of lipids; secretion of cell wall carbohydrates (plant cells); formation of lysosomes; **[1]**

(iv) digest unwanted cell materials; **[1]**

(v) site of protein synthesis; **[1]**

(vi) rough ER involved in protein transport; smooth ER synthesis and transport of lipids/steroids; **[1]**

(d) palisade mesophyll/leaf cell surrounded by cellulose cell wall/animal cell by cell membrane; chloroplasts present in leaf cell/not in animal cell; leaf cell may store starch/storage product in animal cell never starch/is glycogen; leaf cell has large cell vacuole with cell sap; **[3]**

1.4 tissues are collections of similar cells/cells with similar function; usually connected by intercellular material;
epithelial tissues cover internal/line gut/and external surfaces/skin of the body; protect against mechanical damage; connected by basement membrane; in skin consists of many layers; innermost divide/mitosis; outer layers dead/keratinised/ waterproof;
squamous/pavement epithelium single layer of cells; e.g. Bowman's capsule/alveolus/blood capillaries; allow rapid diffusion;
columnar/ciliated epithelium have cilia on surface; beat rhythmically to move mucus in nasal passages/trachea; ovum along oviduct/Fallopian tube;
connective tissue different types of cells in non-cellular matrix; e.g. bone/cartilage; blood is unusual fluid connective tissue; concerned with transport;

muscle tissue, contractile cells;
striated/skeletal/striped muscle necessary for support and locomotion;
smooth muscle in gut/arteries/skin; not under voluntary control; involved in many metabolic functions e.g. nutrition/excretion;
cardiac muscle in the heart; myogenic/beats without being stimulated;
nervous tissue consists of neurones; very specialised for conduction of nervous impulses; occurs in brain/spinal cord; **[25]**

1.5 **(a)** magnify; resolving; close; electrons; magnets; vacuum; transmission; scanning electron microscope; electronmicrograph; **[9]**

(b) light/optical microscope is cheap; portable; relatively easy to use; can view living or dead organisms/tissue; specimens prepared quite simply; Electron microscope very expensive to buy/run; requires special training to use/trained technicians; specimens must be dead; specimens require complex preparation; large/need special room; **[6]**

(c) preparation of the specimen may cause artefacts (distortions) in appearance; may be misleading/making it difficult to interpret image/interpret accurately; **[1]**

1.6 **(a)** to break open cells; **[1]**

(b) water potential of the added solution is the same as the water potential of the tissue and so organelles will not be damaged by losing/gaining water; **[1]**

(c) spinning at high speed so that particles settle out according to their density; as speed increases less dense particles are obtained/ increasing speed get less dense particles; **[2]**

(d) they are too small to be filtered; **[1]**

(e)

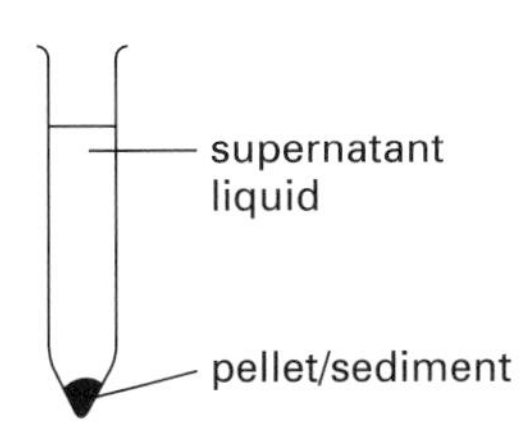

[2]

(f) enzyme/protein; **[1]**

(g) enzymes 'packaged' in organelles may be released and present in homogenate/may start to hydrolyse materials that are required; enzymes work very slowly at low temperatures/damage to the cell fractions will be minimised. **[1]**

(h) nuclei chloroplasts mitochondria ribosomes; **[1]**

1.7 **(a)** 7–8 nm; **[1]**

(b) **A** hydrophilic protein channel; **B** protein; **C** glycoprotein; **D** phospholipid; **[4]**

(c) molecules in the membrane can move around/have a fluid position; proteins are arranged at random/form a mosaic. **[2]**

(d) phospholipid molecules form a double layer/bilayer; polar/hydrophilic heads to outside of membrane; non-polar/ hydrophobic tails towards inside; **[2]**

(e) cholesterol; **[1]**

(f) non-polar/molecules that are not charged; **[1]**

(g) extrinsic protein extends through part of the membrane; intrinsic protein spans the membrane from one side to the other; glycoprotein is a complex between a molecule of carbohydrate and one of protein; **[3]**

(h) form hydrophilic channels/allow polar molecules through; act as carriers/facilitated diffusion/active transport; allow cell recognition; are receptors for hormones on outer cell surface; as antigens; are enzymes/ catalyse reactions at cell membrane; **[3]**

(i) polar molecules are insoluble in cholesterol; **[1]**

1.8 **(a)** to maintain pH/ionic concentration inside and outside the cell; uptake of oxygen/respiratory substrate for respiration; secretion of enzymes; of hormones; excretion of metabolic waste; **[3]**

(b) diffusion – passive; facilitated diffusion – passive; osmosis – passive; active transport – active; endocytosis/pinocytosis/phagocytosis – active; exocytosis – active; **[5]**

(c) the movement of molecules or ions in a liquid or gas, along a concentration gradient/from area of high concentration to an area of low concentration; **[1]**
alveoli; nephron/kidney tubule; ileum; blood capillaries; any living body cells; **[3]**

(d) it is slow; **[1]**

(e) phospholipid/bilayer; hydrophilic channels; facilitated; proteins; receptor/binding; shape; side; kinetic; ATP; passive; **[10]**

1.9 **(a)** permeable to small molecules/water only; **[1]**

(b) A: level of solution in glass tube rises; water moved from high water potential/distilled water; to sucrose solution/lower water potential; by osmosis;
B: level in tubing drops; water moved from high water potential/water in visking tubing; into sucrose solution/lower water potential; by osmosis; **[6]**

(c) Movement is along a concentration gradient; but molecules pass through a differentially permeable membrane; **[1]**

(d) water potential = **B**; solute potential = **A**; pressure potential = **C**; **[3]**

(e) water potential = solute potential + pressure potential
$\psi = \psi_s + \psi_p$ **[2]**

(f) hypotonic solution: low concentration of solute in water; hypertonic solution: has higher concentration of solute in water; isotonic solution: the same concentrations of water and solute; **[3]**

(g) **Delete: (i)** 100; atmospheres; **(ii)** positive; **(iii)** higher; **(iv)** lower; **(v)** F, G; **[6]**

(h) water is hypotonic to red blood cell; water enters cell by osmosis; cell volume increases/cell bursts/haemolysis; saline solution is hypertonic to red blood cell; cell loses water by osmosis; volume decreases/cell membrane becomes crinkled/crenated; **[4]**

(i) **Draw:** a turgid cell; a cell at incipient plasmolysis; a fully plasmolysed cell; **[3]**

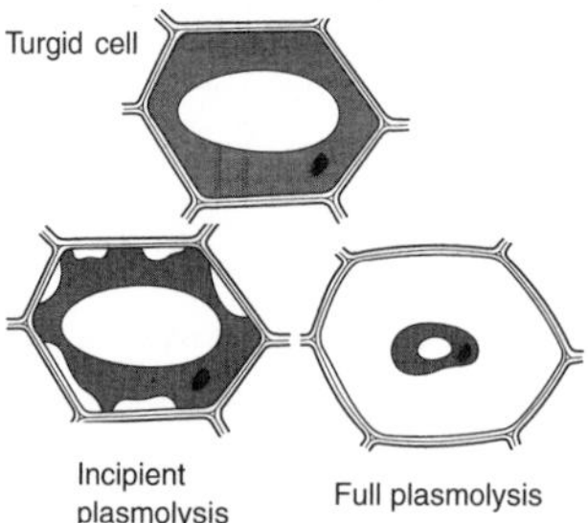

(j) **(i)** X = plasmolysed cell; Y = incipient plasmolysis; Z = turgid cell; **[3]**
(ii) X and Y; **[1]**

1.10 **(a)** **E**; **A**; **C**; **G**; **D**; **F**; **B**; **[7]**

(b) **(i)** in nerve fibres; involved in the action potential/restoration of resting potential;
(ii) inner mitochondrial membranes; ATP synthesis; chemiosmotic theory; **[4]**

(c)

ENDOCYTOSIS

Nutrients attach to receptors on cell membrane

Vacuole containing undigested food moves to cell membrane

Cell membrane folds in enclosing nutrients in a vacuole where digestion occurs

Fuses with membrane. Waste leaves the cell

EXOCYTOSIS

[4]

(d) all three use energy from ATP; **[2]**

1.11

	Active transport	Osmosis	Facilitated diffusion	Diffusion
Occurs in living cells only	✓	✗	✓	✗
Particles travel from high to low concentration	✗	✓	✓	✓
Uses protein carriers	✓	✗	✓	✗
ATP provides the source of energy	✓	✗	✗	✗

[4]

1.12 **(a)** the elderly/the very young/people with weak immune systems/on chemotherapy/with HIV/with heart disease/diabetics/asthmatics; **[3]**

(b) coat of protein; nucleic acid core DNA or RNA; no nucleus; no membranes/organelles; **[4]**

(c) only nucleic acid present is RNA; Human Immunodeficiency Virus (HIV); **[2]**

(d) virus mutates; vaccine may not be effective against the mutated strain; **[1]**

(e) **Plasmid:** small circular piece of DNA/into which genetic material can be inserted; common in bacterial cells; **Virion:** a virus particle; **[2]**

(f) by mutation/of viral RNA; **[1]**

(g) the number of virus particles inhaled is too few to cause disease symptoms; number of virus particles must be increased by replication; **[2]**

2 Biological molecules

2.1 **(a)** B **[1]**

(b) molecule is unequally charged; hydrogen slightly positive/oxygen slightly negative; **[1]**

(c)

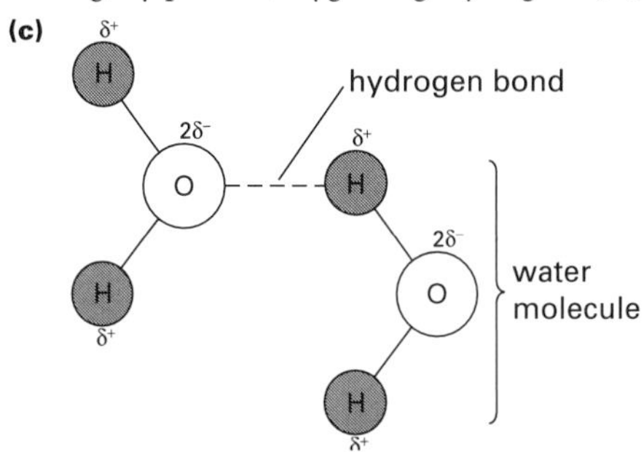

the water molecules are difficult to separate; **[3]**

(d) sodium chloride; glucose; amino acids; **[3]**

(e) charges on water molecule attract to ions of polar molecule; weakens bonds between ions of polar molecule; ions separate and dissolve; **[2]**

(f) **(i)** high; **(ii)** contract; high; **(iii)** more; false; 4°C; the densest water does not freeze, but sinks to the bottom of the lake/pond where organisms can survive winter; **(iv)** transparent; photosynthesis; **(v)** evaporation; high; water is using latent heat (energy) from the body to evaporate, which cools the person; **[12]**

2.2 Most abundant molecule on Earth/human body 60% water; metabolic reactions take place in aqueous medium; hydrolysis of polymers to monomers e.g. proteins to amino acids; diffusion, respiratory surfaces must be moist; for gaseous exchange/absorption of glucose or amino acids through ileum to blood circulation; important solvent; e.g. blood/tissue fluid/lymph transport dissolved substances; metabolic waste e.g. urea, excess salt, excreted in urine; many body secretions high in water e.g. digestive juices/semen/tears; important lubricant e.g. mucus in gut/vagina/nasal passages; synovial fluid in synovial joints reduces friction/pleural fluid in thorax; protective/support, amniotic fluid for fetus; support in eye, vitreous and aqueous humour; thermoregulation: sweating to cool body/maintain correct temperature; raw material photosynthesis/basis of all food chains/replenishes oxygen in atmosphere; sea and freshwater are environments for various human foods e.g. fish/seaweeds/can be farmed; in polar regions ice can be used as building material (igloos); **[10]**

2.3 **(a)** **A** making bonds between monomers and giving out a water molecule; linking of glucose to form starch; **B** breaking of bonds between monomer units of a polymer; by the addition of water; proteins broken down to amino acids; **C** basic unit which can be used to build complex molecules; glucose; **D** complex molecule made from many linked monomers; starch/lipid/protein; **E** synthesis of a complex organic molecule requiring energy; photosynthesis; **F** breaking down of a molecule with release of energy; respiration; **[12]**

(b)

	Polysaccharides	Lipids	Proteins
Elements present in all molecules	carbon, hydrogen, oxygen	carbon, hydrogen, oxygen	carbon, hydrogen, oxygen, nitrogen
Subunits	hexose sugars	triglycerides (fatty acids +glycerol)	amino acids
Bonds between subunits	glycosidic	ester	peptide
Are they found in cell surface membranes?	yes	yes	yes
Function	energy source, structure	energy source	growth, repair
Are they stored in human tissues?	yes (as glycogen)	yes (as fat)	no
Can they act as a human respiratory substrate?	yes (converted first to glucose)	yes (converted to fatty acids)	yes (but only when human is starving)

[14]

2.4 **(a)** crystalline solids; soluble in water; sweet; **[3]**

(b) 3 carbon atoms – trioses; 5 carbon atoms – pentoses; 6 carbon atoms – hexoses; **[3]**

(i) ribose; $C_5H_{10}O_5$;

(ii) deoxyribose; $C_5H_{10}O_4$; **[4]**

The deoxyribose molecule contains one less oxygen atom than ribose; **[1]**

(c) one or more chemical compounds with the same chemical formula/contain exactly the same type and number of atoms in a molecule; **[1]** glucose, fructose and galactose all have chemical formula $C_6H_{12}O_6$; **[1]**

(d) glucose and fructose are structural isomers; both have the same chemical formula; but the atoms are linked differently in the two ring structures; **[3]**

(e) sucrose + water = glucose + fructose

$C_{12}H_{22}O_{11} + H_2O = C_6H_{12}O_6 + C_6H_{12}O_6$ **[2]**

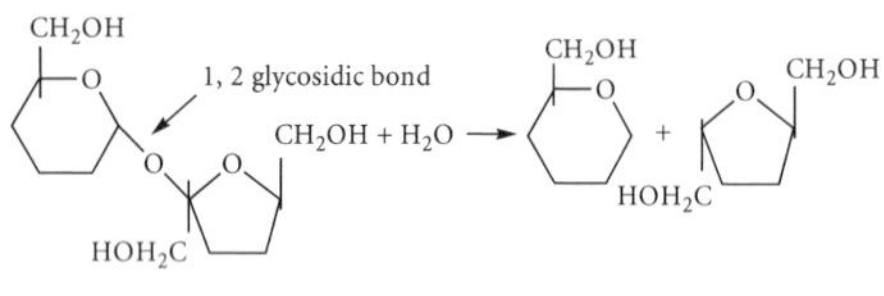

sucrose + water ⟶ glucose + fructose **[3]**

(f) (i) A; (ii) B; (iii) C; F; H; (iv) E; (v) J; E; (vi) B; E; G; (vii) J; (viii) H; (ix) D; J; (x) C; F; G; **[18]**

2.5 (a) glucose; galactose **[1]**

(b) fructose **[1]**

(c) all of them **[1]**

(d) maltose and lactose produce red precipitate; sucrose, no change; **[3]**

maltose and lactose are reducing sugars; **[2]**

(e) (i) Benedict's solution remains blue; **[1]**

(ii) the acid has hydrolysed the glycosidic bond between the two monosaccharide subunits (monomers); **[1]**

all monosaccharides are reducing sugars/ reducing sugar is now present/reducing sugars produce brick red precipitate on heating with Benedict's solution; **[1]**

2.6	Cellulose	Glycogen	Sucrose	Starch
Monomer	β - glucose	α - glucose	α - glucose	α - glucose
Number of subunits	many	many	two	many (form amylose and amylopectin)
Are side chains present?	no	yes	no	yes
Type of bond(s) between monomers	β 1-4 glycosidic bonds	α 1-4 glycosidic bonds	α 1-4 glycosidic bonds	α 1-4 glycosidic bonds α 1-6 glycosidic bonds
Is it soluble in water?	no	no	yes	no
Where does it occur naturally in living cells?	in plant cell walls	in human liver and muscle cells	in sugar beet and sugar cane; transported through plants in phloem	in plant cells
Function	structural (cell walls)	storage carbohydrate	transport carbohydrate also storage	storage carbohydrate

[**Total 7** 1 mark for each totally correct row]

2.7

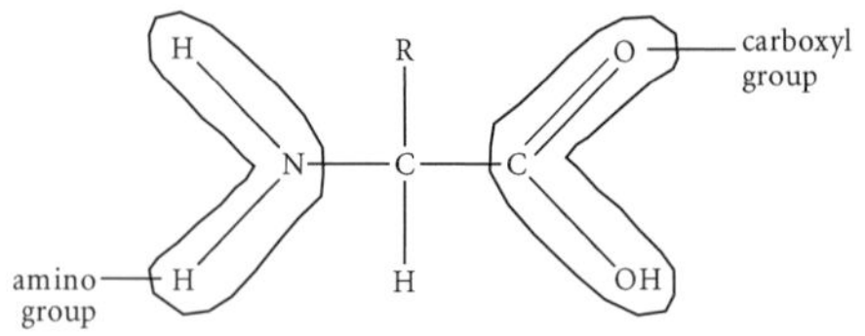

(a) Correct labels. **[2]**

R indicates the radicle of the amino acid; **[1]**

(b) condensation reaction; **[1]**

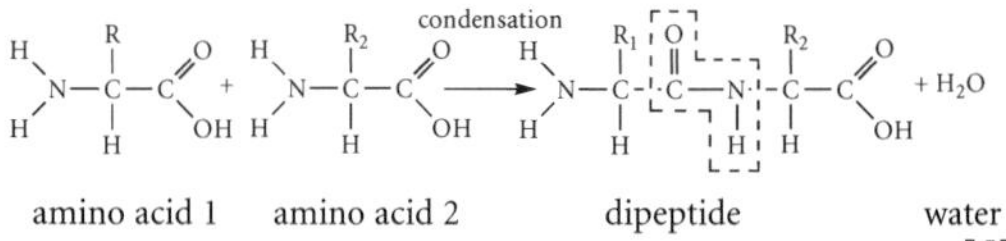

amino acid 1 + amino acid 2 ⟶ dipeptide + water **[4]**

(c) polypeptide (chain); **[1]**

(d) Essential amino acids must be provided in the diet; Non-essential amino acids can be synthesised in the body (transamination); **[2]**

(e) pH; basic; amphoteric; dipolar; gain; alkaline; lose; negatively; isoelectric point; **[9]**

(f) blood (pH must be maintained 7.35–7.45); **[1]**

(g) about 20; **[1]**

2.8 (a) (i) D; (ii) B; (iii) C; (iv) A; **[4]**

(c) globular: soluble in water; polypeptide chain folded; complex tertiary structure; **fibrous:** little/no tertiary structure; tough; insoluble in water; high tensile strength; polypeptide chains with cross links; **[8]**

(d) fibrous: collagen; keratin; **globular:** remainder; **[7]**

2.9 (a)

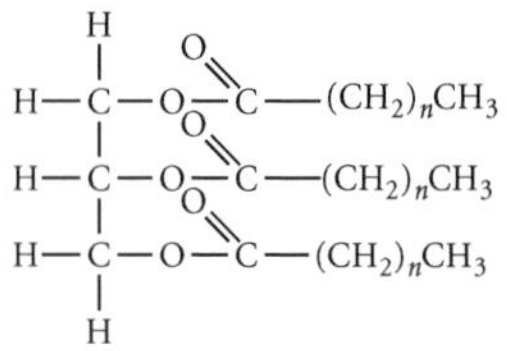

[4]

(b) condensation; **[1]**

(c) saturated fatty acid molecule, all bonds are single; unsaturated fatty acid molecule contains a double bond; **[1]**

(d) lipids; fatty acids; glycerol; solid; liquid; water; organic solvents; ethanol (or another suitable example); (complex) alcohols; hair; **[10]**

(e) a phosphate group condenses with one of the –OH groups in the glycerol component of the triglyceride; **[1]**

instead of a fatty acid; **[1]**

(f) hydrophilic = polar/soluble in water; [1]
hydrophobic = non polar/insoluble in water/soluble in organic solvents; [1]

(g) component of cell membranes; synthesis of acetylcholine; [2]

(h) liver; any two from testosterone/oestrogen/progesterone/aldosterone;; [3]

2.10 Trigycerides compact molecules; insoluble in water; can be stored in high concentration (in cells); they store twice the amount of energy as an equivalent mass of carbohydrate; and yield twice the water from respiration; energy is released only by aerobic respiration; insulate animals (in cold climates)/hibernating animals; play a part in thermoregulation; give bouyancy to aquatic animals; stored as oils in seeds and fruits which may be important foods; give rise to waxes e.g.cutin for waterproofing; give rise to phospholipids/vital to cell membranes; also sterols e.g cholesterol; [10]

3 Enzymes

3.1 (a) cells; globular proteins; tertiary; speed up; used up;/chemically changed; (these 2 points can be in either order) specific; optimum; denatured; inhibitors; [10]

(b)

Enzyme	Metabolic process involved in	Function
decarboxylase	Krebs cycle	removal of carbon dioxide;
hydrolase;	digestion;	hydrolysis
DNA ligase	DNA synthesis;	formation of new bonds;
oxidase;	electron transfer chain	oxidation;
dehydrogenase;	Krebs cycle;	removal of hydrogen

[9]

3.2 (a) heating; [1]

(b) activation energy/energy necessary for reaction to occur in absence of enzyme; [1]

(c) activation energy/energy needed for reaction when enzyme present; [1]

(d) any correct examples e.g. catalase; hydrogen peroxide = water + oxygen/$2H_2O = 2H_2O_2 + O_2$; sucrase; sucrose = glucose + fructose; (The reaction must be correct for the enzyme named for each second point) [4]

(e) any correct example e.g. endopeptidase; breaks bonds between amino acids in polypeptide chains; exopeptidase; breaks bond at end of polypeptide chain; [4]
(The reaction must be correct for the enzyme named for each second point.)

(f) enzymes allow reactions to occur/faster/at low temperatures; must be present to catalyse vital reaction; [1]

(g) genes [1]

3.3 (a) Lock and key: enzyme has an active site where substrate molecule can bind; active site is 3–12 amino acids/(which form) a specific shape; enzyme is lock and substrate is key which fits it; enzyme-substrate complex is temporary structure; enzyme released for use again/product(s) formed; Induced fit suggests enzyme structure is more flexible; active site wraps around enzyme-substrate complex; changing conformation of enzyme;
causes substrate to break into products;
(You get one mark for good quality in your written answer and a mark for each specified term used correctly and in a logical order.) [8]

(b) X-ray diffraction. (Compare shape of enzyme alone with shape of enzyme-substrate complex.) [1]

3.4 (a) substance which slows down or stops an enzyme controlled reaction; [1]

(b) structure/shape/property of active site; by blocking reactions can confirm pathways in a metabolic pathway; [2]

(c) (i) competitive inhibitor/inhibition; [1]
(ii) irreversible/non-competitive inhibitor/inhibition; [1]

(d) malonate is competitive inhibitor to succinate; for active site on succinate dehydrogenase/enzyme in Krebs cycle; found in mitochondrial matrix; slows down aerobic respiration; [3]

(e) (i) known volume of substrate and of enzyme added; at known/constant temperature; need method of detecting that reaction is proceeding e.g. colour change; measure accurately time for reaction to finish; repeat several times/average results; repeat with different volumes over suitable range of substrate concentrations; [4]
(ii) line E; [1]

(iii) at high substrate concentration E shows rate of reaction similar to the control/W/no added inhibitor; because there are many substrate molecules to compete with inhibitor; for enzyme active site; inhibitor has less effect on rate of reaction; rate at high substrate concentration for test D is much lower than W/no inhibitor added/control; inhibitor affects enzyme permanently/ denatures enzyme; there are fewer enzyme molecules to react with substrate; **[4]**

(f) non-protein substance/example; **[1]**
essential for enzyme to function; **[1]**

3.5 **(a)** Quote two temperatures from graph in answer; at low temperature (5°C) rate of reaction is very slow; as temperature rises molecules gain more (kinetic) energy/move faster; number of collisions between molecules of enzyme and substrate increases; at (40°C) rate of reaction is maximum/is optimum temperature for this enzyme; as temperature rises higher molecules vibrate more; hydrogen bonds break; tertiary structure of protein/enzyme changed permanently; enzyme/protein denatured; enzyme can no longer bind with (distorted) active site; enzyme-substrate complex can no longer form; rate of reaction declines rapidly; **[8]**

(b) For temperatures up to 40°C; a rise in temperature of 10°C; will double the rate of reaction; **[3]**

or $Q_{10} = \frac{\text{rate of reaction at (n +10°C)}}{\text{rate of reaction at n°C}}$

3.6 **(a)** the concentration of hydrogen ions/H^+; measure of acidity of a solution; **[2]**

(b) to keep the pH constant/from changing significantly; **[1]**

(c) mass of liver/tissue; volume of buffer solution; temperature; size of manometer tubing; **[3]**

(d) pH 5 = 0.2 pH 6 = 0.8 pH 7 = 5.4 pH 8 = 6.7 pH 9 = 10.0 **[all correct 2][4 correct 1]**

(e) Rate on vertical axes and pH on horizontal axis; axes labelled; suitable scale (graph should fill most of paper); correctly plotted points; points joined correctly with ruler; **[4]**

(f) optimum pH fell between 7 and 8/extra measurement gives more accurate result; **[1]**

(g) indicate extra point and use (different type/dashed) line to show change this has on curve;; **[2]**

(h) 7.6 (accept 7.4–7.8); **[1]**

(i) hydrogen bonds had broken; tertiary structure of catalase/(globular) protein was changed; active site changed so substrate/hydrogen peroxide could not bind with it; **[2]**

(j) use one pH close to optimum; choose range of temperatures/15–60°C; suitable number of temperatures/4–6; **[2]**

3.7 **(a)** Figure 3.5, rate of reaction increases initially and then becomes stationary; increasing substrate concentration does not increase rate of reaction; all active sites on enzyme are occupied; enzyme concentration limits rate of reaction; Figure 3.6, rate increases linearly/in proportion to increasing concentration of enzyme; there are no limiting factors; **[4]**

(b) substrate concentration decreases/all substrate is used; **[1]**

(c) something which slows down rate of chemical reaction; **[1]** temperature; low molecules have low kinetic energy/fewer collisions; if too high will denature enzymes and slow down rate of enzyme controlled reaction; pH; too high or too low can interfere with enzyme conformation/shape; low substrate concentration or low enzyme concentration; reduces enzyme-substrate complex formation; light (light dependent reactions only); may not provide sufficient energy for reaction/intensity too high may destroy chlorophyll (plants); **[6]**
[1 mark for limiting factor and 1 mark reason]

3.8 **(a)** adsorption on to an insoluble matrix; (covalent) binding to cellulose/ceramics/ nylon; entrapment within a gel; encapsulation in a selectively permeable membrane; **[any 3]**

(b) enzyme can be used again/many times; does not contaminate the end product; does not pollute environment; may have increased thermostability/less heat sensitive/not denatured at 65°C; may function in wider range of conditions/in organic solvents; **[any 3]**

(c) untreated milk contains lactose; no glucose, galactose; **or** treated milk has no lactose; contains glucose, galactose; **[2]**

(d) heat with Benedict's solution; untreated milk remains blue/shows no reducing sugars/glucose, galactose; treated milk gives red precipitate/reducing sugars present; **[2]**
test both samples with glucostix/diastix/equivalent; glucose present in treated milk only; **[2]**

(e) fast flow rate lactose will not be hydrolysed; enzyme does not have time to bind with lactose; slow flow rate galactose accumulates in column; inhibits action of lactase on lactose; **[4]**

(f) babies secrete lactase; as they get older; those genetically lactose intolerant ; secrete less/no lactase; **[2]**

(g) all carbohydrate in whole milk is lactose/lactolite <0.24 g lactose/provamel only 0.1 g sugar/0.6 g carbohydrate; glucose in lactolite/not in whole milk; soya milk has almost half the energy of the other two/2.1 g fat compared with 4 g; more fibre in soya milk; same calcium in all; **[any 3]**

(h) calcium deficiency; **[1]**

(i) people from different races intermarry; gene pool increases; gene for lactose intolerance will become more frequent in NW European population; lactose intolerance likely to become more common; **[3]**

(j) human milk: more carbohydrate/lactose; less protein; less calcium; more fat; **[2]**

3.9 **(a)**

Apples chopped and crushed/pressed
↓
Pectinase added
↓
Pulp filtered to collect juice
↓
Pectinase added
↓
Juice put into cartons or bottles **[5]**

(b) in the cell walls/middle lamellae; **[1]**

(c) pectin is hydrolysed/broken down; into soluble products; **[2]**

(d) break open the cells/release the juice; increased surface area for pectinase to act on; **[1]**

(e) increase the yield/volume of juice; decrease the viscosity of the juice; to clarify/make it clear/less cloudy; **[3]**

(f) cellulase; it would break down cellulose in the cell walls; allowing greater yield of juice; **[2]**

(g) how juicy/yield of juice; colour of juice; taste of juice; cost of apples; **[3]**

4 Breathing and gas exchange

4.1 **(a)** 4.1B cuboidal epithelium; **[1]**

(b) smooth muscle; **[1]**

(c) 4.1C ciliated epithelium; **[1]**

(d) cilia beat towards pharynx; moving mucus; and trapped dirt away from lungs; **[3]**

(e) 4.1A squamous epithelium (thinnest layer); **[1]**

(f) thin (small diffusing distance); **[1]**
permeable; **[1]**
moist (since chemical reactions in the cell happen in solution, gases dissolve at the gas exchange surface); **[1]**

4.2 warmed; moisture/water vapour/humidity; trachea; air pressure; bronchioles; surface area; surfactant; surface tension; **[8]**

4.3 **(a)** A: 13.28 kPa; **[1]**
B: 5.41 kPa; **[1]**

(b) higher partial pressure in alveolus than blood; oxygen diffuses along diffusion gradient from higher concentration/partial pressure to lower concentration/partial pressure; **[2]**

(c) lower because it is being used up during respiration; **[1]**

(d) CO_2 produced during cellular respiration in tissues; diffuses from cells into blood; removed from blood by diffusion into alveoli maintaining gradient; **[3]**

4.4 **(a)** the larger the surface area the greater the rate of diffusion as more molecules may contact the surface at any point in time; the greater the difference in concentration the greater the diffusion gradient, so diffusion happens faster; the thinner the membrane the faster the rate of diffusion because there is less diffusing distance; **[3]**

(b) solubility in water; size; temperature **[3]**

4.5 breathing movements change the volume of chest cavity; volume changes cause air pressure changes; creates pressure difference between the inside of the thorax and atmosphere; air moves from higher pressure to lower pressure area; intercostal muscles raise ribs; and diaphragm lowers; during inhalation causing volume increase; and pressure decrease inside thorax; intercostal muscles and diaphragm relax; and elastic recoil of muscle fibres occurs; causing volume decrease in thorax; and pressure increase; during exhalation (The order of these points is not crucial as long as they are set out logically) **[Any 8]**

4.6 **(a)** kPa; **[1]**

(b) pressure inside lungs/thorax is less than outside/atmospheric pressure; **[1]**

(c) causes muscle fibres to return to original length during exhalation decreasing volume in thorax; **[1]**

(d) lungs are not fixed to the inside of thorax; pleural membranes line inside of thorax and run over outer surface of lungs; puncture means air moves into space between thorax and lungs; no pressure difference between inside of thorax and atmosphere outside; [must include last point, but any two others for a total of 3] **[3]**

4.7 **(a)** tidal volume is volume of air exchanged at each resting breath; ventilation rate is tidal volume times the number of breaths/min (total amount exchanged min^{-1}); **[2]**

(b) frequency of inhalation $min^{-1} = 102/10 = 10.2$; ventilation rate = tidal volume × frequency of inhalation/min;
$= 450 \times 10.2$
$= 4.59\ l\ min^{-1}$; **[3]**

(c) increase in demand for oxygen; increased cellular respiration; increase in activity/exercise; **[1]**

(d) volume of air left in lungs after strongest possible exhalation; **[1]**

4.8 **(a)** time/s **[1]**

(b) tidal volume is the volume of air entering and leaving the lungs; at each natural resting; breath; **[2]**

(c) vital capacity **[1]**

(d) residual volume is the amount of air left in the lungs after the strongest possible exhalation/cannot be breathed out **[1]**

(e) $1.2\,dm^3$ **[1]**

4.9 **(a)** A area: $0.01 \times 0.01 \times 6 = 0.0006\ cm^2$; **[1]**
A volume: $0.01 \times 0.01 \times 0.01 = 0.000001\ cm^3$; **[1]**
B area: $10 \times 10 \times 6 = 600\ cm^2$; **[1]**
B volume: $10 \times 10 \times 10 = 1000\ cm^3$; **[1]**

(b) $A = 0.0006/0.000001 = 600$; **[1]**
$B = 600/1000 = 0.06$; **[1]**

(c) B; **[1]**

(d) the lower the surface area to volume ratio (as B) the more need for mass flow; as diffusion alone cannot provide enough oxygen; **[2]**

4.10 **(a)** smooth muscle in airways contracts; airways become narrower; **[2]**
(inflammation may be mentioned but other two points required)

(b) wheeziness due to severe breathing difficulty; **[1]**

(c) allergens; diesel residues; fur; dust (especially from dust mites); cigarette smoke; viral infections; cold air temperatures; anxiety; **[any 2]**

(d) managing own environment/lifestyle (e.g. don't smoke; change bedding to low allergen type, etc); use medication (preventers, relievers)/inhalers; monitor peak flow/response to own lifestyle situations to recognise triggers; **[3]**

4.11 **(a)** from evening to early morning/1800–0800 h; **[2]**

(b) person is tired/requires rest; **[1]**

(c) raises ventilation rate/litres of air breathed per minute; improves ventilation over a period of time; **[2]**

(d) nebuliser use means that a greater dosage of medicine is given, which should not be sustained for too long as side effects might arise over a longer period; **[1]**

4.12 **(a)** 1: fetus; **[1]**
2: cilia stop moving/reduced surface area for gas exchange; **[1]**

(b) nicotine; **[1]**

(c) carbon monoxide; **[1]**

(d) a substance which causes cancer; **[1]**

(e) pass in bloodstream of mother; through placenta; into bloodstream of fetus affecting fetal cells; **[3]**

(f) organ systems may not be fully developed; more vulnerable to disease/infection; **[2]**

4.13 **(a)** increase quite sharp between 1950 and 1960; increase slow between 1960 and 1970; decrease quite rapid between 1970 and 2000; **[3]**

(b) in 1950 after second world war women were more used to working outside home; were more independent; had more money available; increase in advertising aimed at women; society accepted women smoking/became more liberal; **[any 3]**

(c) similar in that 4.6 first increases and then decreases; increase in mortality rate continues for longer than increase in smoking rate; due to time lag between starting to smoke and developing cancer; lung cancer decreases after 1980, later than smoking rate decreases; **[any 3]**

(d) lifelong non-smoker has the lowest rate of developing lung cancer; lifelong smoker has the highest rate of developing lung cancer; the younger a smoker stops, the faster their risk of developing cancer drops; **[3]**

5 Blood and circulation

5.1 1 sodium, potassium, calcium, magnesium, chloride, hydrogen carbonate, hydrogen phosphate, sulphate (any one, or another suitable);
2 glucose, amino acids, urea, lipids (any one or another suitable);
3 albumens, globulins, fibrins, or antibodies, hormones, etc. (any one);
4 and 5 functions of plasma: acts as a solvent; distribution/transport; role in thermal buffering/thermoregulation; pH buffer (any two);;
6 and 7 granulocytes, agranulocytes, or any two names e.g. leucocytes: neutrophils, eosinophils, basophils, monocytes, lymphocytes;;
8 and 9 features of white blood cells include can change cell shape, can leave blood capillary to enter tissues, may be phagocytic, may produce histamine, heparin, antibodies, etc. (any 2 appropriate);;
10 and 11 features of erythrocytes: disc shaped, lack nucleus, large surface area, contain haemoglobin, about 5 million per mm^3 (any 2);;
12 and 13 functions erythrocytes: transport O_2 and CO_2;;
14 fragments of megakaryocyte cells;
15 more;
16 function of platelets: clotting; **[Total 16]**

5.2 **(a)** check health of donor e.g. not lacking Hb; check for communicable diseases; **[1]**

(b) standard deviation is a measure of the variation of a set of values from the arithmetic mean; the value of calculating standard deviation is that it gives an indication of the spread of values from the mean e.g. small standard deviation means that the values will be very similar to the mean, not much variance; **[2]**

(c) 125g/l of Hb is very low for males; below normal health threshold so may be experiencing health problems; **[1]**

(d) 2.6%; **[1]**

(e) 86.9%; **[1]**

5.3 **(a)** 1 globular protein; globulin sub-unit; globin; **[1]**
2 metal ion/iron II ion/Fe ion; **[1]**
3 haem unit; **[1]**

(b) each Hb carries 8 oxygen atoms;
$8 \times 250 = 2000$ oxygen atoms; **[2]**

(c) insufficiently soluble; **[1]**

(d) carboxyhaemoglobin; **[1]**

(e) anaemia; **[1]**

5.4 **(a)** loading tension for Hb is measured at 95% since complete saturation is never achieved, and unloading tension is measured at 50% as Hb never gives up all its oxygen; **[1]**
loading tension = 11.20 kPa
unloading tension = 3.5 kPa
difference = 11.20 – 3.5 = 7.7 kPa;
[1 for correct unit used]

(b) S shape occurs because the first oxygen molecule to bind to Hb changes the shape of the Hb molecule, so that subsequent oxygen molecules bind more easily; **[2]**

(c) sketch second curve to right of original, but should start and finish in same positions;
[1 for reasonable accuracy]

(d) Hb has less affinity for oxygen as partial pressure of CO_2 rises; hence oxygen unloads more rapidly in respiring tissues; **[2]**

5.5	Feature			
Disorder	Is a genetic disorder	Iron content is low	Low blood sugar occurs	Oxygen carriage is reduced
Anaemia	✓ and ✗	✓	✗	✓
Haemophilia	✓	✗	✗	✗
Sickle cell	✓	✗	✗	✓
Hyperglycaemia	✓ and ✗	✗	✗	✗

[4]

5.6 arteries and veins 3-layered wall; middle smooth muscle layer; thicker muscle in arteries than veins; arteries carry blood away from heart; have to withstand high blood pressure; and can stretch; and have elastic recoil; helping to pump blood; veins at lower pressure; contain valves; to assist unidirectional flow; take blood back to heart from organs; arteries, veins and capillaries lined by endothelium; capillaries one layered; small size; close proximity to cells; allow easy diffusion; retain plasma proteins; **[12]**

5.7 capillary: **B**; **E**; artery: **C**; **G**; **H**; vein: **A**; **D**; **F**; **[8]**

5.8 **(a)** higher the pressure the higher the rate of flow, higher the cross-section the higher the rate of flow; higher cross-section means lower pressure, so speed of flow depends on both factors; **[2]**

(b) different sized vessels occur; some tissues are further from the heart than others; higher speed might cause damage to tissues; slower flow is desirable to allow sufficient exchange of materials; **[2]**

5.9 **(a)** Refer to *Understanding Advanced Human Biology* or other standard texts for this information, which is readily available, using 25 separate points of information. **[25]**

(b) within the lower thorax; between the lungs; **[2]**

(c) to act as a lubricant to reduce friction when heart moves; **[1]**

(d) to provide oxygenated blood to cardiac muscle; **[1]**

5.10 pump; fibres; linked; planes or directions; intercalated discs; respiration; myogenic; **[7]**

5.11 **(a)** order is A C D B **[2]**

(b) A and C; **[1]**

(c) increased pressure of blood in the ventricles as they fill; **[1]**

(d) to ensure blood flow does not return to atria as ventricles contact; **[1]**

(e) higher in main arteries than ventricles; **[1]**

5.12 **(a)** nodes marked correctly on diagram; (sinoatrial node situated near where the venae cavae enters right atrium, atrioventricular node also located in right atrium but nearer entrance to ventricle) **[1]**

(b) would disrupt heart cycle; normally initiates heart beat; sends impulses to AV node; sets heart rate; **[4]**

(c) perkinje/perkyne/Bundles of His; which run down septum to apex of heart; and through walls of ventricles; **[2]**

(d) delays wave of excitation slightly; passes impulses on to Bundle of His; **[2]**

(e) correctly sketched **[3]**

5.13 **(a)**

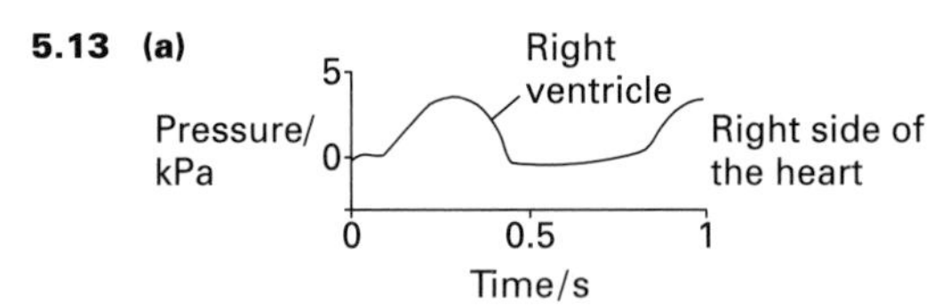

(1 point if curve is generally much lower but follows similar pattern, second point for accuracy) **[2]**

(b) pressure is higher in the left ventricle because the lumen is smaller; and the walls are thicker muscle which contract with more force; **[2]**

(c) X is towards the tops of the steep downward gradient of the curve, on the right hand side of the peak; **[1]**

(d) contraction of the atrium; **[1]**

5.14 **(a)** electrocardiogram; **[1]**

(b) SA node intitiates electrical activity; which spreads down across atria towards AV node; as atria contract **[2]**

(c) after contraction the muscle cells relax; and the membrane potentials return to the resting potential; **[2]**

5.15 **(a)** different antigens on blood cells; agglutination possible; **[1]**

(b) fetal blood passes to placenta in umbilical artery; close contact between maternal blood and fetal capillaries; diffusion; fetal organ systems not fully functional before birth; **[4]**

(c) connects two sides of fetal heart; redirects deoxygenated blood to aorta and then to placenta; [2]
(d) happens via ductus arteriosus; pulmonary circulation is reduced; because lungs are not functional; [3]
(e) foramen ovale and ductus arteriosus seal off; blood flow to fetal liver increases; [2]

5.16 (a) forms part of the haemoglobin molecule needed for oxygen carriage; [1]
(b) they were more used to the nature of the test/not so nervous – or any sensible suggestion; [1]
(c) significantly; increases performance in IQ test; [2]
(d) contains no active ingredient; [1]
(e) brain/mental activity requires energy; so oxygen required, which is carried by haemoglobin; [2]

6 The body and exercise

6.1 (a) breathe faster; heart rate increases; feel hotter; sweat; surface vasodilation; [4]
(b) respiratory system; cardiovascular system; skin; (as appropriate) [2]
(c) 50% of 170 = 85 and 75% of 170 = 127.5 i.e. 85–127.5 [1]
(d) Exercising at less than target heart rate does not produce as much benefit/stimulate cardiac development; exercising at more than target heart rate might stress body systems more than is desirable; [2]
(e) physiology of body changes with age or fitter/stronger when younger; [1]

6.2 (a) 6 minutes; [1]
(b) greater muscle activity; higher rate of respiration; [2]
(c) oxygen deficit/lack of oxygen; anaerobic respiration; pyruvate respired to lactate; lactate/lactic acid builds up in muscles; [2]
(d) amount of oxygen required; to completely respire the accumulated lactate; [2]
(e) lactic acid toxic; accumulates in muscles causing pain/cramp; [2]

6.3 (a) volume of blood pumped from ventricles at each heart beat; [1]
(b) stroke volume = cardiac output/heart rate; [1]
(c) $0.105 \times 66 = 6.93\ dm^3$ [2] (including correct unit)
resting; [1]

6.4 (a) hindbrain; carotid arteries; aorta; [2]
(b) increase in CO_2 concentration/pH drop; [1]
(c) inspiratory centre; [1]
(d) stretch receptors; [1]
(e) expiratory centre; [1]
(f) prevent hyperventilation; damage to lungs by over inflation; [1]
(g) drugs; illness; injury; [2]

6.5 (a) heart rate controlled by reflex actions; of autonomic nervous system; stretch receptors in walls of aortic arch; carotid arteries; and vena cava; these affect inhibitory centre in brain; slow heart rate; chemoreceptors; sensitive to concentration CO_2/pH; stimulate acceleratory centre; speeding up heart rate; SA and AV nodes and heart muscle; affected by impulses from cardiovascular control centre as described above; [10]
(b) adrenaline increases heart rate; thyroxine indirect effect via metabolic rate changes; nerve endings release neurotransmitters which act as chemical coordinators; [3]

6.6 (a) skin; cardiac muscle; skeletal muscle; [any 2]
(b) additional flow is required for skeletal muscles/muscles require additional oxygen/glucose supplies; [1]
(c) gut; [1]
(b) athlete has increased bulk of cardiac muscle, stroke volume is higher than in unfit individuals; therefore heart rate does not need to increase as much to achieve increased cardiac output; [2]

6.7 respiration; myoglobin; anaerobic; fat/lipid; [4]

6.8 (a) 85% of its resting length [1]
(b) at 105% of its resting length [1]
(c) more cross-bridges between myosin and actin; so more force exerted; [2]
(d) regular exercise at close to maximum force of contraction; [1]

7 Cell division

7.1 microscope; chromatin; DNA; histones; denser; chromosomes; chromatids; centromere; **[8]**

7.2 **(a)** **A** telophase; **B** anaphase; **C** metaphase; **D** interphase; **E** prophase; **[5]**
correct order: interphase, prophase, metaphase, anaphase, telophase, interphase; (**Note You may put interphase first or last.**) **[1]**

(b) they will be identical/have the same DNA/ have same number of chromosomes; they are identical to the parent cell; **[2]**

(c) chromatids would not separate and both would go to one daughter cell/one daughter cell would be one chromosome short/a mutation occurs; **[1]**

(d) You will find suitable diagrams in your text. Note the instructions to use 6 chromosomes. You will get one mark for each drawing and one for the labelling. There is no need to label structures such as the spindle more than once. **[8]**

7.3 **(a)** 1 = metaphase 2; 2 = (late) prophase 1; 3 = telophase 2; 4 = anaphase 1; **[4]**

(b) spindle pole; **[1]**

(c) the spindle; microtubules/tubulin/protein fibres; **[2]**

(d) chiasma crossing over/genetic material is exchanged; leads to variation/new varieties; allows adaption to environmental change; **[3]**

(e) pair of homologous chromosomes/ bivalent; **[1]**

(f) cell division includes complete division of nucleus and cytoplasm; nuclear division is division of nucleus only; **[2]**

7.4 **(a)** anaphase 1; **[1]**

(b) The centromeres have divided and the chromatids separated. One from each chromosome is pulled to opposite spindle poles.

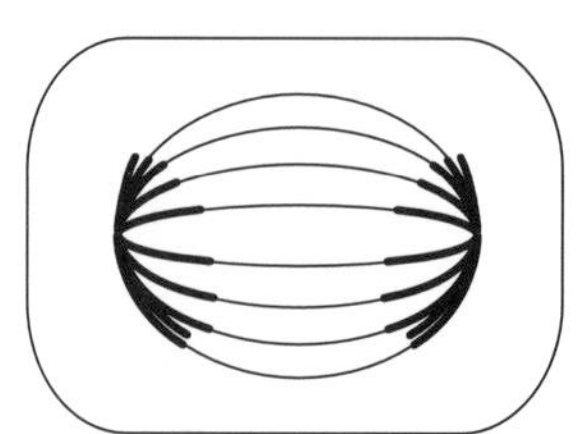

[4]

7.5

	Mitosis	Meiosis 1
Chiasmata form and genetic variation may result	✘	✓
Chromatids separate during anaphase	✓	✘
Paired homologous chromosomes line up across the spindle equator	✘	✓
Chromatids become visible during prophase	✓	✓
The number of chromosomes is halved	✘	✓
Condensation of chromosomes occurs	✓	✓

[6]

7.6 **(a)**; **(d)**; **(f)**; **(h)**; are all mitosis; **(b)**; **(c)**; **(e)**; **(g)**; **(i)**; **(j)**; are meiosis **[10]**

7.7 **(a)** haploid cells have a single set of chromosomes/n; diploid cells have a pair of each chromosome 2n; polyploid cells have multiple sets of chromosomes; **[3]**

(b) T, W; **[1]**

(c) 3; **[1]**

(d) W; **[1]** we are told that it came from T, and it is identical to T; **[1]** T might also have been produced by mitosis but it could have been formed by the fusion of two cells, each with 3 chromosomes.

(e) X Y; They have half the number of chromosomes of parent cell T; **[2]**

(f) two copies of A in cell W/one copy in X and Y; chromosomes in Y and X are different to one another/different to those in W; crossing over occurred/chiasmata formed during prophase 1 as X and Y were being produced; the chromatids became tangled/broken; genetic material was exchanged; **[4]**

(g) Any four from the following

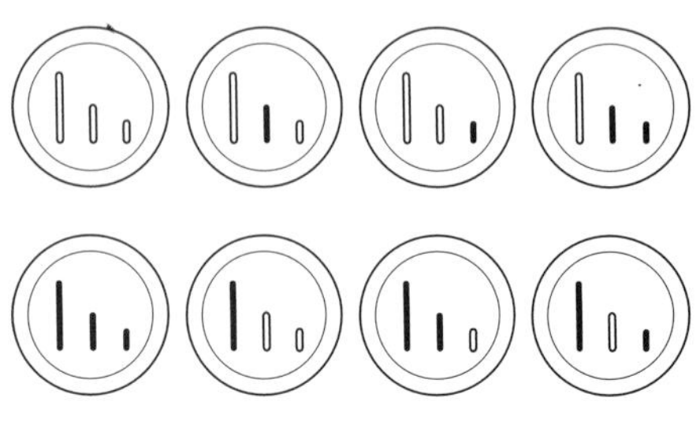

[4]

(h) testes/testis; ovary/ovaries; **[2]**

(i) **A** 46; **B** 46; **C** 46; **D** 23; **E** 0; **F** 46; **G** 46; **H** 23; **[8]**

7.8 (a) M; the DNA is halved/half DNA goes to each new cell; **[2]**

(b) S; the curve shows quantity of DNA rises **[1]**

(c) cell grows/enlarges; **[1]**

(d) interphase; **[1]**

(e) striped muscle cell/smooth muscle cell/liver cell. There are many examples of specialised cell you could choose. **[1]**

(f) at the start of G1 each chromosome is a single chromatid; chromatids are DNA; before a cell can divide a second chromatid must be replicated so that the parent cell can pass identical genetic material to the two new cells; **[2]**

(g) normal p53 causes the destruction of abnormal/mutated cells; if p53 is absent abnormal/cancer cells divide freely; cancer cells without p53 divide indefinitely; **[2]**

8 DNA and protein synthesis

8.1 (a) A = sugar/pentose sugar/deoxyribose; B = phosphate; C = hydrogen bond; D = guanine; E = thymine; **[5]**

(b)

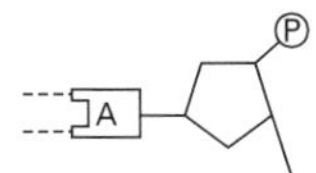

[1]

(c) adenine and guanine are purine bases **[2]**

(d) thymine and cytosine are pyrimidines **[2]**

(e) purines pair with pyrimidines; adenine pairs with thymine/guanine pairs with cytosine; **[2]**

8.2 (a) the DNA molecule remains intact and instructs another copy to be formed; (How this might happen was never explained) **[1]**

(b) **F A E G D C B** or **F E G D A C B** **[7]**

The DNA molecule unwinds from one end and the strands separate as hydrogen bonds between complementary bases break. The reactions are catalysed by DNA polymerase enzyme. Nitrogenous bases are exposed along each original strand. They act as templates for free nucleotides present in the nuclear sap. These nucleotides join together by condensation reactions to form two new identical DNA strands. Each new DNA molecule consists of a strand from the original molecule and a new complementary strand. This process is called 'semi-conservative replication'.

8.3 (a) ribonucleic acid; nucleotides; ribose; (adenine; guanine; cytosine; uracil; in any order) (mRNA/messenger RNA; tRNA/transfer RNA; rRNA/ribosomal RNA; in any order) **[10]**

(b) it has one more oxygen atom per molecule;

(c) **[1]**

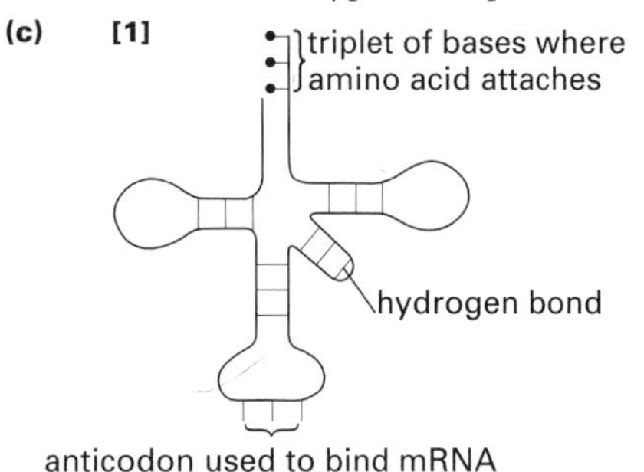

[3]

8.4

	DNA	RNA
Monomer units	nucleotide	nucleotide
Purine bases	adenine guanine	adenine guanine
Pyrimidine bases	thymine cytosine	uracil cytosine
Pentose sugar	deoxyribose	ribose
Shape of molecule	double helix	single helix
Molecular mass	high	lower
Chemical stability	very stable	variable/mRNA transient
Where found in cell	nucleus	nucleus and cytoplasm
Types of this nucleic acid	one	three/mRNA, tRNA, rRNA

[9]

8.5 (a) sequence of codons/triplets of bases; on DNA; each codes for an amino acid; in polypeptide; **[3]**

(b) **(i)** more triplets than there are amino acids; **(ii)** three bases that code for an amino acid; **(iii)** an individual base is part of one triplet only/ can only be read once; **(iv)** a triplet codes for the same amino acid in all living organisms; **[4]**

(c) **(i)** intron: non-coding DNA; exon DNA codes for amino acids;

(ii) codon: triplet of bases on DNA/mRNA; anticodon: sequence of bases on tRNA complementary to codons on mRNA; **[4]**

(d) U A C G U G A A A G U G C G A U U C; **[1]**

(e) tyrosine,valine, lysine, valine, arginine, leucine; **[1]**

8.6 **(a)** hydogen; template; transcribed; mRNA; polymerase; nucleus; cytoplasm; ribosome; amino acids; anticodon; complementary; ribosome; condenses; amino; polypeptide; ATP; **[16]**

(b) **(i)** is hydrolysed/breaks up to nucleotides; **(ii)** can be re-used; **(iii)** assumes tertiary structure; **[3]**

(c) a group of ribosomes; that translate one strand of mRNA; so forming a number of identical polypeptides; **[2]**

(d) triplex: prevent translation; DNA strands cannot unwind to expose bases; to act as template; **[2]**

antisense: prevent transcription; tRNA anticodons; cannot pair with mRNA; **[2]**

9 Gene technology

9.1 **(a)** Circular strands of DNA; breaking open cells and centrifugation; **[2]**

(b) staggered cut; at specific sequence on DNA; exposes unpaired bases; complementary to cut in plasmid DNA; **[3]**

hydrogen bonds; **[1]**

(c) making bond between gene and plasmid DNA permanent/secure; **[1]**

(d) exposing bacteria to short high temperatures; calcium ions/charge; **[2]**

(e)

Letter	Name of enzyme	Function of enzyme
V	reverse transcriptase	translation of single strand of cDNA
W	DNA polymerase	synthesises second strand of cDNA
X	restriction endonuclease	cuts DNA at specific base sequence
Y	DNA ligase	forms permanent bond in recombinant DNA

[8]

(f) A G G T G T C T A A A C;
T C C A C A G A T T T G; **[2]**

(g) plasmids multiply; bacteria multiply; required product is synthesised by recombinant DNA; using metabolism of bacteria; **[3]**

(h) insulin; human growth hormone; or any other correct answer; **[2]**

9.2 **(a)** gene in plasmid which allows for the selection of plasmids carrying desired recombinant DNA sequence; **[1]**

(b) **(i)** antibiotics; **[1]**

(ii) Q; **[1]**

(c) **(i)** one bacterium/bacterial cell; **[1]**

(ii) 3, 4, 6 ; **[1]**

colonies growing on B are resistant to tetracycline; they do not carry the gene for X; as it inactivates the gene for tetracycline resistance; /or vice versa **[3]**

(iii) inoculate a sterile; nutrient medium; with cells from colonies that grew on plate A but not on plate B; incubate at suitable temperature; **[3]**

9.3 **(a)** a short length of radioactive/ chemiluminescent; single stranded DNA; that can bind to a complementary sequence on a short length of DNA; **[2]**

used when a sequence of bases has been identified; **[1]**

(b) autoradiography – **G**; hybridisation – **F**; electrophoresis – **C**; **[3]**

9.4 **(a)** high level of synthesis of products/gene switched on much of time; **[1]**

(b) cells containing recombinant DNA/DNA from another organism; **[1]**

(c) more food/opines; more cells to inhabit; **[2]**

(d) resistance to herbicides; immunity to viruses/other plant pathogens; unpalatable to insect pests; slower ripening/increased shelf life for fruit and vegetables; **[3]**

(e) marker gene; e.g. antibiotic resistance; **[2]**

(f) inserted gene is only expressed by cells containing it/must be present in all cells needing the characteristic; **[1]**

(g) isolation of plastid from *A. tumefasciens*; using restrictase enzymes; plastids cut open with restriction enzyme; required gene sequence removed; inserted into plastid; annealed; with DNA ligase; **[5]** (in logical order)

(h) *Agrobacterium tumefasciens* carrying recombinant DNA plasmids; **[1]**

(i) micropropagation; **[1]**

(j) totally sterile conditions; **[1]**

(k) cereals are main food source of world population; *A. tumefasciens* cannot be used to transform/change DNA of cereals; this technology not available to huge populations on this planet; **[2]**

(l) more plants/cereals can be transformed; **[1]**

9.5 **(a)** allows rapid amplification; of small sequence of DNA; **[2]**

(b) free nucleotides; DNA sequence; (oliginucleotide) primer; polymerase enzyme; **[4]**

(c)

Stage	1	2	3
Process	denaturation	annealing	polymerisation
Approximate time	30–35 sec	20 sec	30 sec
Temperature °C	90–95	55	75
What happens	original DNA separates into 2 strands	primers bind to separated DNA	2 DNA double helices form

[8]

(d) **(i)** 2; **(ii)** 2^{10}; **(iii)** 2^{20}; **[3]**

(e) this enzyme was denatured by the high temperatures; **[1]**

(f) thermophiles/bacteria living in hot spring/ named species; enzymes can be obtained by genetic engineering; **[2]**

(g) use in forensic work from small samples DNA left at crime scenes; Human Genome project; prenatal testing for genetic disorders; on specimens from history especially bones; in evolutionary research looking for relationships between species; any other relevant points. **[4]**

9.6 **(a)** **(i)** non-coding region of DNA; consisting of many short base sequences repeated many times/minisatellites; **[2]**

(ii) DNA that codes for a protein; **[1]**

(iii) unique pattern of DNA banding/introns and exons; **[1]**

(b) Any three: to match DNA from a crime scene with DNA of a suspect; to establish relationships between individuals/paternity/ family members/origins of protected species; evolution/how closely different species or fossils and living species are related; **[3]**

(c) **B**; **D**; **A**; **C**; **[4]**

(d) **(i)** C and F; **[2]**

(ii) A and E; **[2]**

(e) **(i)** D and B could not be siblings; because they do not share sufficient sequences on their DNA fingerprint; **[2]**

(ii) B and F could be siblings; because they both have bands from individuals A and E, who are the parents of B; **[2]**

9.7 ethical objections to artificial manipulation of genetic material; named example; gene for resistance to herbicide can be inserted into crop; example resistance to 'Roundup' in soya beans used quite widely in USA; herbicide can be sprayed at any time to kill weeds; biodiversity is reduced/fewer plant species; leads to fewer insects/bees; fewer birds/small mammals; herbicide residues may enter food chain and reach humans. Companies manufacturing herbicides will make huge profits; GM crop in one field could easily cross pollinate/ contaminate unmodified crop in next/or wild plants; genetic modification spreading to weeds may lead to resistance to herbicides in weeds/'superweed'; will need more herbicide/ more toxic herbicide to control them; difficult to guarantee organic crops; genetic markers used during genetic modification may be for antibiotic resistance; these genes may survive digestion; concentrate higher in food chain; **[8]**

10 Parasites, pathogens and disease

10.1

	Tape-worm	Liver fluke	Malaria	Sleeping sickness
Name of parasite	Taenia	Fasciola	Plasmodium	Trypanosoma
Name of vector if one occurs	no vector	no vector	mosquito/ anopheles	Tsetse fly
Name of secondary host if there is one	pig	water snail	none	none
Symptoms caused in human host	abdominal pain; distention; flatulence; nausea	digestive disturbance; liver damage	fever; shivering; pain in joints; headache; vomiting; convulsions; coma	can be chronic or acute: fever; weakness, aching, swollen lymph glands; mental confusion; lethargy; coma

[20]

10.2 **(a)** surface protein coat, protein core; containing single stranded RNA; **[3]**

(b) by inclusion in the membrane and emptying inwards; **[1]**

(c) allows production of double stranded DNA from RNA; which can integrate with host DNA; **[2]**

(d) transcription; **[1]**

(e) translation; **[1]**

(f) disruption at the host cell surface; cell death; **[2]**

(g) not recognised as antigenic; as it is hidden within lymphocyte; **[2]**

(h) weakens it as lymphocytes are destroyed; **[1]**

(i) protect host surface cell membrane from penetration by virions; inhibit reverse transcriptase enzyme; prevent assembly of new virion particles; prevent penetration of new host cell; other similar point; **[3]**

10.3 **(a)** Line A: 1, 2, 4, 8, 16, 32, 64, 128, 256, 512, 1024 **[1]**

(b) Line B: 0.000; 0.301; 0.602; 1.204; 1.505; 1.806; 2.107; 2.408; 2.709; 3.010 **[1]**

(c) should be a straight line graph, points plotted accurately and scales set out accurately **[5]**

(d) logs allow large and small population sizes to be plotted on the same graph; logs reflects the true rate of population growth irrespective of populations size, and slope equals rate of growth; **[either point 1]**

(e) as there is a doubling at each generation; the log of the population number increases in direct proportion to time; **[2]**

10.4 **(a)** bacteria cultured from all samples taken; less bacteria were cultured from the skin which had been towel dried; **[2]**

(b) washing method was constant so is unlikely to make a significant difference to the results; drying method was different so was likely to have made a difference to the results; **[2]**

(c) paper towel removes more bacteria; paper towel removes surface layers of skin cells which carry bacteria; paper more effective at drying; people dry more carefully with paper towel; people do not wait for hot air blower to dry completely; moisture remains on skin which is air dried, so more bacteria collect on swab; **[any 2]**

(d) the site where the samples were collected; that the sample was random e.g. first 50 people to use an office bathroom facility; nature of paper towel e.g. not impregnated with disinfectant etc; method of swabbing and culturing; **[any 2]**

10.5 reduce populations of mosquitos which carry malaria by use of pesticides; limit places where mosquitoes can breed; introduce natural predator of mosquito; use of mosquito nets; insect repellants; develop a vaccine to give people immunity to Plasmodium; health education; **[5]**

10.6 **(a)** reduction in vector; **[1]**

(b) incidence of malaria fell in areas where spraying occurred; **[1]**

(c) not applied sufficiently widely; to eradicate malaria; **[2]**

(d) aimed to control disease by diagnosis; and treatment; rather than prevention of infection by controlling vector; **[3]**

(e) residues of insecticide which do not break down easily; remain in soil/waterways or animals in food chain; **[2]**

(f) might be effective elsewhere; if adequately applied; **[2]**

10.7 **(a)** prevented infection; easy/cheap to use; live virus spreads to unvaccinated people; **[3]**

(b) no more disease/saves lives/saves illness; less money spent on treatment/vaccination; **[2]**

(c) OPV contains live virus but Salk contains dead virus; OPV is taken by mouth while Salk is injected; **[2]**
(d) weakened viruses still in the population could mutate to more dangerous forms; **[1]**
(e) increase the pool of viruses as OPV vaccine is live **[1]**

10.8 (a) Mycobacterium tuberculosis **[1]**
(b) antibiotics; such as streptomycin **[1]**
(c) survived exposure to antibiotics and development resistance **[2]**

10.9 breach surface barrier; through break in skin; or by active penetration mechanism; natural entry ports such as pores and hair follicles; breathed in and enter through respiratory membranes; eye surface/conjuctiva and cornea; anal and genital membranes; might be ingested in food; **[5]**

10.10 (a) increases; greater in males than females; **[2]**
(b) yes; in each case there is a lower Hb value for people with intestinal parasite than those without; **[2]**
(c) half the people are likely to fall within this range (indicates reference values for normal population); and lower and upper 25% groups are at the extremes of the whole sample; **[1]**
(d) females of menstrual age lose Hb regularly; so expected values are lower; **[2]**
(e) Peruvians live at high altitude; so higher Hb level expected; **[2]**

10.11 (a) (i) being infected by/becoming a carrier/picking up an organism; **[1]**
(ii) the microorganism being present in personal microflora/an organism you carry on your body surface; **[1]**
(b) student has new lifestyle; new social group; greater social interaction; greater exposure to microbes; changes to diet; **[5]**
(c) rapid increase in carriage rate of *N.meninigitidis*; tailing off slightly towards end of week; **[2]**
(d) rate increases with increased visits to bar; **[2]**
(e) being a smoker; being in a self-catered hall of residence; being in a male only hall of residence; **[3]**
(f) statistical techniques give an objective view and additional comparative information from raw data; **[1]**
(g) 95% of the sample will be expected to fall within that range; **[1]**

10.12 (a) notifiable diseases are potentially life-threatening; spread rapidly; cause severe damage to organ systems; cannot be easily treated/cured; might help indicate a particular source of disease that affects many people; able to offer information about patterns of disease and lifestyle factors associated with disease; **[3]** e.g. of disease which is not notifiable is: cold; 'flu; chest infection; sore throat; athlete's foot; etc. **[1]**
(b) bacterial: by ingestion in food or drink; viral: by the respiratory route in vapour droplets or from eating shellfish; toxins (chemical or bacterial in origin); **[3]**
(c) more people eat in restaurants; there are more retail food outlets; fewer people self-cater; more highly processed foods with shorter shelf life are available; similar answer; **[2]**
(d) some people do not vaccinate children for religious or cultural reasons; some people enter the UK while the disease is developing within them (during incubation period); **[2]**
(e) makes people aware that immunisation is available; makes the dangers of a disease known; removes fears about the safety of the immunisation itself; **[2]**
(f) smallpox is eradicated globally; rabies has been controlled by a strict policy on quarantine and/or checks on all animals entering the UK; **[2]**
(g) cholera is not prevalent in the UK so immunisation is not required; **[1]**
(h) immunisation recommended for people going on holiday to a tropical country/a country where cholera is not controlled; **[1]**

11 Heart disease

11.1 (a) (i) high blood pressure; **[1]**
(ii) blockage of blood supply to brain, causing death of some brain tissue; **[2]**
(iii) arteries supplying heart muscle become blocked reducing blood supply; **[2]**
(b) living in an industrial/urban environment may involve being less physically active, eating more processed food; and smoking/other drug use; **[3]**
(c) males; **[1]**

11.2 1: lowers water potential of blood drawing water into plasma, increasing hydrostatic pressure; **[1]**
2: nicotine is a stimulant acting on smooth muscle in small arteries and arterioles, causing contraction and increasing pressure within them; **[1]**
3: genetic; **[1]**
4: puts a strain on the heart and affects fitness; **[1]**

11.3 **(a)** chest pain caused by restricted blood flow to cardiac muscle; **[1]**
(b) fatty deposit in blood vessel; **[1]**
(c) a blood clot in circulating blood; **[1]**
(d) haemorrhage or blockage of an artery in the brain causing death of an area of brain cells and resultant loss of function; **[1]**
(e) surgery to bypass a blocked coronary supply; **[1]**
(f) patch on a blood vessel wall where swelling or a burst may occur; **[1]**

11.4 **(a)** they become over-stretched and lose flexibility/elasticity; **[1]**
(b) loss of elasticity of arteries causes increased resistance to blood flow; and heart has to work harder to pump blood; heart muscle mass may increase; enlarged heart muscle; **[2]**
(c) glomerular capillaries affected and filtration is less efficient; **[1]**
(d) they may burst; **[1]**

11.5 **(a)** a thrombus; may originate in the heart as circulation in the coronary supply is slowed; and may pass to the brain causing a blockage; and stroke; **[2]**
(b) that increased fish consumption leads to decrease in death from heart attack; **[1]**
(c) other dietary considerations such as saturated fat/salt consumption; exercise habits; smoking; genetic factors; obesity; **[5]**
(d) high in unsaturated fatty acids so do not contribute to atheroma formation; **[2]**

11.6 **(a)** avoid a high fat diet/eat foods high in polyunsaturated fatty acids to avoid encouraging cholesterol formation; avoid high salt; exercise regularly and moderately; do not smoke; do not be overweight/obese; **[5]**
(b) less loss of working days/more people working; so smaller loss to treasury of contributions/more contributions generated; lower cost of treatment; funding for healthcare can be directed elsewhere; health of nation is improved; **[4]**
(c) circulatory problems caused by ductus arteriosus/hole in heart; or valve malfunction; since lifestyle factors are probably less important; e.g. most children exercise as a part of play, and are non-smokers etc.; **[3]**

11.7 **B** beta-blocker; **C** diuretic; **A** calcium channel blocker; **[3]**

11.8 **(a)** lack of donor; cost; **[2]**
(b) more have coronary heart disease; **[1]**
(c) survival rate decreases with time; although the rate of death slightly decreases; **[2]**
(d) rejection by the immune system; **[1]**

11.9 **(a)** electrocardiogram/ECG; **[1]**
(b) muscle contraction initiated at SA/sinu-atrial node;
spreads across atria to AV/atrioventricular node; **[2]**
(c) peak should be marked between 0.4 and 0.55; **[1]**
(d) heart muscle rests/relaxes; actin/myosin/myofilaments return to resting position; diastole; **[3]**

11.10 **(a)** right ventricle 4 kPa;
right atrium 1 kPa; **[2]**
(b) Ventricle has thicker muscle; and exerts greater force of contraction/pressure; **[2]**
(c) carries blood to the lungs; **[1]**
(d) pressure is greatest in ventricle when it contracts/systole; which happens after atrium has emptied/has relaxed; **[2]**

12 Immunity, diagnosis and disease

12.1 **(a)** a particle or organism which is recognised as non-self by human body, causing an immune response; **[1]**
(b) clumping together of small particles to form larger masses; **[1]**
(c) type of white blood cell that responds to antigens; **[1]**
(d) medicine which slows the growth of or kills bacteria; **[1]**
(e) antibiotic which retards the growth of bacteria; **[1]**

(f) tracking the progress/spread of disease and associated conditions/causative factors; **[1]**

(g) globular proteins which make up antibodies; **[1]**

12.2 antigens; lymphocytes; leaving/squeezing out of; cytoplasm; engulf/take in; membrane bound; antigen/contents/bacterium; **[7]**

12.3 **(a)** epidermis; composed of cornified layer; granular layer; and malpighian layer; then deeper dermis; **[5]**

(b) dermis contains connective tissue including collagen; splits disrupt protective barrier provided by skin to prevent infection by organisms; **[2]**

(c) strong physical barrier; made of many layers of cells which become keratinised; dead cells; as they reach the surface; also waterproof; so infective agents cannot pass through easily; **[3]**

(d) 3.9 litres = 3900 cm^3
3900 × 2.47 × 100/48 000 = 20.07% **[3]**
(includes workings, correct answer and % sign with answer)

12.4

Feature	T-lymphocyte	B-lymphocyte
Produced from stem cells which arise in bone marrow	✓	✓
Help agglutinate antigens	✗	✓
Engulf antigenic particles by phagocytosis	✓	✗
Produce antibodies	✗	✓
Bind to antigens, destroying them	✓	✗
Form plasma clone cells	✗	✓

[12]

12.5 **(a)** role is to produce lots of antibody rapidly, and once the antigen is destroyed there is no need for the antibody; (unless infection re-occurs) **[1]**

(b) each antigen is different; has specific protein markers; recognised by B-lymphocytes; antibody molecule matches a particular antigen marker; and binds to it, as a key fits a specific lock; **[3]**

(c) gives protection against secondary infection/subsequent exposures to antigens; and allows rapid immune response; by rapid production of antibody; **[2]**

(d) **(i)** IgM; **[1]**

(ii) IgM produced to a similar level as the first exposure; IgG produced to higher level simultaneously; **[2]**

12.6 **(a)** antibodies produced in response to exposure to antigen e.g. sore throat; **[1]**

(b) colostrum provided in breast milk to suckling infant; **[1]**

(c) an immunisation using antigens that will not cause disease e.g. Rubella vaccination; **[1]**

(d) injecting ready made antibodies in emergency situation e.g. tetanus antiserum; **[1]**

12.7 immunisation involves using harmless forms of an antigen to cause an immune response; developed to protect against serious infectious diseases; such as smallpox, tuberculosis, poliomyelitis or other example; has allowed eradication of some diseases such as smallpox; and drastically reduced prevalence of many infectious diseases; and reduced mortality rate due to infectious diseases; increasing average lifespan; and reducing debilitation due to disease; antibiotics fight infections; which are already established and are used to speed recovery; and prevent health deteriorating; quickly reducing symptoms; and damage from infection; reducing absence from work due to sickness; raising level of health of population generally; **[10]**

12.8 **(a)** alkaline phosphatase; **[1]**

(b) destroyed by liver; **[1]**

(c) continuing damage to/disease of an organ; **[1]**

(d) **(i)** 48 hours; **[1]**

(ii) 2 × t½ for feline lactate dehydrogenase is 60 hours; hours and 2 × t½ for canine lactate dehydrogenase is 36 hours; so feline enzyme takes 14 hours longer; **[2]** (including showing workings and unit (hours))

(e) **(i)** molecular mass of proteins means they are too large to pass through capillary wall; so not filtered in glomerulus/Bowman's capsule/kidney; **[2]**

(ii) albumen might pass through glomerulus if it is malfunctioning; and so level in plasma would drop; **[2]**

(iii) the amino group from excess proteins; **[1]**

(iv) if not filtering correctly; less urea might pass through glomerulus; more might pass out of collecting ducts into interstitial tissue; more might be reabsorbed into vasa recta that pass to renal vein; **[2]**

(v) albumen and total protein levels; do not decrease in plasma; in urinary tract obstruction; **[2]**

(vi) insufficient data to be conclusive; need to confirm diagnosis; or any relevant answer; **[1]**

12.9 antibodies produced by body; during immune response to infection; antibodies can be linked to enzymes; when antibody recognises antigen (say in blood sample) it binds to it; the attached enzyme acts like an indicator; bringing about a colour change; or other observable effect; rapid method of detecting specific microbial infections; **[5]**

13 Inheritance

13.1 **(a)** chromosome: structure of DNA and protein found in nucleus; consisting of two chromatids held by centromere; number constant for every species; **[2]**
gene: sequence of DNA coding for a polypeptide; unit of inheritance; **[1]**
allele: alternative forms of a gene; dominant or recessive; **[1]**

(b) dominant gene: expressed in phenotype when a single allele only is present in genotype; **[1]**
co-dominant gene: two alleles each contributing to phenotype when present at same locus in genotype; gives intermediate/different expression from dominant or recessive alleles; **[1]**

(c) homozygote: has two identical alleles/both dominant or both recessive occupying given gene locus on homologous chromosomes; heterozygote: has dominant and recessive alleles at given gene locus on homologous chromosomes; **[2]**

(d) monohybrid inheritance: of a single pair of alleles; dihybrid inheritance: of two unlinked alleles; **[2]**

(e) round; all F_1 plants had round fruit; if either parent has a dominant allele it will show in F_1; **[2]**

(f) **(i)** RR
(ii) rr
(iii) Rr **[3]**

(g) F_1 parent → (columns); F_1 ↓ (rows)

gametes	R	r
R	RR round fruit	Rr round fruit
r	Rr round fruit	rr oval fruit

correct gametes; **[1]** any 2 correct genotypes; **[1]** phenotypes that match genotypes **[1 each]** **[max.4]**

(h) 50% are heterozygous (Rr); **[1]**

13.2 **(a)** Both parents are carriers for PKU/genotype Pp

mother's gametes (columns); father's gametes (rows)

	P	p
P	PP normal	Pp normal
p	pP normal	pp has PKU

correct gametes for both parents; **[1]**
4 correct genotypes; **[1]** 4 phenotypes correctly match genotypes; **[1]**

(b) 0.25/25%/1 in 4; **[1]**

(c) PP; Pp/pP; **[2]**

(d) 0.33/33%/⅓ of healthy/normal children have genotype PP; 0.66/66%/⅔ of healthy children are carriers Pp; **[2]**

(e) PKU parent carrier has genotype Pp and produces gametes P and p: normal partner has genotype PP, all gametes P;

gametes	P	p
P	PP normal	Pp normal

[1]

Probability of a child with PKU is 0/zero; **[1]**

(f) heel pricked/to get blood sample; tested for phenylpyruvic acid; **[2]**

(g) make abnormal thick mucus; causes severe congestion of lungs; and pancreatic duct; interferes with digestion; requires daily physiotherapy; medication; reduces life expectancy; **[4]**

(h) medication; physiotherapy; **[2]**

(i) gene therapy; **[1]**

13.3 **(a)** 1; 2; **[2]**
(b) 4; **[1]**
(c) dominant; **[1]**
(d) appears in every generation/does not skip generations; **[1]**
(e)

Parent 5

	gametes	d	d
Parent 8	D	Dd dwarf	Dd dwarf
	d	dd normal	dd normal

correct gametes for parents ; correct genotypes; correct phenotypes; **[3]**
(f) 0.5/$\frac{1}{2}$/50% chance of dwarfism; **[1]**
(g) no; **[1]**

13.4 **(a)** two alleles occupying same locus; on homologous chromosomes; are both expressed in phenotype; **[2]**
(b)

Phenotype	Genotype	Antigen(s) on red blood cells
MN;	MN	M and N;
N;	NN;	N
M;	MM;	M;

[6]

(c)

gametes	M	N
M	MM	MN
N	MN	NN

Children's phenotypes from two parents MN are MM:2MN: NN **[3]**
(d) **(i)** group A; **(ii)** group AB; **(iii)** group B; **[3]**
(e) O: I^o I^o; A: I^o I^A and I^A I^A; B: I^o I^B and I^B I^B; AB: I^A I^B; **[4]**
(f) O = universal donors can give blood to any other group safely/red cells carry no antigens so are not attacked by patient's immune system; AB can receive blood from all other groups; **[2]**
(g) 1 = group O or A; 2 = group B; 3 = group B; 4 = group AB; **[4]**
(h) RR for genotype of red parent and rr for genotype of white parent; red flower gametes carried R; white flower gametes were r; F_1 plants had genotype Rr; phenotype was pink because R and r are co-dominant; both are expressed when together in genotype; gametes from F_1 plants were 50% R and 50% r; fertilised at random;

gametes	R	r
R	RR red	Rr pink
r	Rr pink	rr white

table; shows 1 red:2 pink:1 white plant in F_2 generation; actual plants grown were approximately true to this prediction; **[8]**

13.5 **(a)** **(i)** tt; tt^b; **(ii)** T^At^b; T^At; **[4]**
(b) 1 = T^At^b; 2 = T^At^b; 3 = tt^b; 4 = T^AT^A; 5 = T^At^b; 6 = t^bt^b; **[6]**
(c)

Cat 1

	gametes	T^A	t^b
Cat 5	T^A	T^AT^A ticked	t^bT^A spotted
	t^b	t^bT^A spotted	t^bt^b classic

correct gametes; **[1]** correct genotypes; **[2]** correct phenotypes; **[2]** **[max 4]**
(d) they are siblings/inbreeding; undesirable recessive characteristics could show in phenotype; **[1]**
(e) it must be bred with a classic cat/t^bt^b; a test cross; if his cat is carrying a mackerel allele some/half of the kittens will be mackerel; if his cat is not carrying mackerel some/half kittens will be classic; **[4]**
(The easiest way to answer this is to include diagrams of possible outcomes of the cross.)
(f) genotype of Y would be T^At; crossed with mackerel tt he would have the chance of half mackerel; and half spotted tabby kittens;

Y

	gametes	T^A	t
mackerel	t	T^At spotted	tt mackerel
	t	T^At spotted	tt mackerel

[3]
(g) all kittens must have a dominant agouti allele; without it the markings are not expressed; if both parents are heterozygous for recessive agouti allele kittens may be bred homozygous for recessive agouti; **[2]**

13.6 **(a)** inheritance of 2 pairs; of different characteristics; determined by unlinked genes/genes on different chromosomes; **[2]**
(b) **(i)** ccHH; **(ii)** CChh; **(iii)** CcHh; **[3]**
(c) white flowers; hairy stems; **[2]**
(d) 9:3:3:1; **[1]** orange flowers/smooth stems: orange flowers/hairy stems:white flowers/ smooth stems:white flowers/hairy stems; **[1]**

(e) orange/smooth

orange smooth

gametes	CH	Ch	cH	ch
CH	CCHH orange smooth	CCHh orange smooth	CcHH orange smooth	CcHh orange smooth
Ch	CCHh orange smooth	Cchh orange hairy	CcHh orange smooth	Cchh orange hairy
cH	CcHH orange smooth	CcHh orange smooth	ccHH white smooth	ccHh white smooth
ch	CcHh orange smooth	Cchh orange hairy	ccHh white smooth	cchh white hairy

correct gametes for both parents **[1]**, all genotypes correct **[2]** (2 incorrect **[1]**), all phenotypes correct **[2]**, quality of diagram good **[1]**, prediction confirmed **[1]**

(f) there would be a high proportion of plants with same phenotypes as the original parents/orange smooth and white hairy; fewer new combinations/orange hairy and white smooth; **[2]**

(g) one parent would have stamens removed before maturity; acts as female parent; flower enclosed in polythene bag; to prevent pollination by insects; pollen from other/male parent would be transferred using brush/cotton bud; to stigma of demasculated flower; cross would be carried out in reverse; seeds carefully collected and labelled; germinated; so results can be collected; **[4]**

13.7

	Dominant allele(s)	Recessive allele(s)
The probability of heterozygous parents having a child expressing this allele is 1 in 4	✗	✓
A mother is not affected by this sex-linked allele, but all her sons who inherit it from her are	✗	✓
F_1 plants from a cross between different homozygous parents have the phenotype for this allele	✓	✗
Neither parent is affected by this allele, but some of their children are	✗	✓
This allele is always expressed in a heterozygous phenotype	✓	✗
In a dihybrid cross 9/16 of the F_2 generation show these characteristics in their phenotype	✓	✗
Huntington's disease is an example	✓	✗
This allele is expressed as colour blindness	✗	✓

You must have the whole row correct to gain each mark **[8]**

13.8 **(a)** alleles/genes; Y; pair/partner/allele; phenotype; recessive; XY; sex-linked; homogametic; reversed; XX; **[10]**

(b) **(i)** female normal wings; **(ii)** male white eyes; **(iii)** female red eyes; **(iv)** male normal wings; **[4]**

(c) **(i)** X^RY; **(ii)** X^NX^N and X^NX^n; **(iii)** X^rX^r; **(iv)** X^nY; **[4]**

(d) **(i)** red eyed female (X^RX^r)

white eyed male (X^rY)

gametes	X^R	X^r
X^r	X^RX^r red eyed female	X^rX^r white eyed female
Y	X^RY red eyed male	X^rY white eyed male

25% red eyed female:25% white eyed female:25% white eyed male:25% red eyed male

[1 for each correct: gametes; genotypes; phenotypes; probability; 4]

(d) **(ii)** female miniature wings (X^nX^n)

male normal wings (X^NY)

gametes	X^n	X^n
X^N	X^NX^n female normal wings	X^NX^n female normal wings
Y	X^nY male miniature wings	X^nY male miniature wings

50% (all females) have normal wings/are carriers; 50% (all males) have miniature wings; **[1 for each correct: gametes; genotypes; phenotypes; probability; 4]**

(e) **(i)** 1 = X^HX^h; 7 = X^HX^h; 8 = X^HY; 10 = X^hY; 11 = X^HX^h; **[5]**

(ii) 1 4 7 11 16 17 **[3]**

[1 mark for each 2 correct]

(f) males with haemophilia have mothers who are carriers; more males/all sufferers are male in this tree/female haemophiliacs rare;

male haemophiliacs have normal sons; have carrier daughters; **[any 2]**

13.9 gametes from blue parent Bd and b^1d; gametes from cinnamon parent b^1D and b^1d; **[2]**
draw Punnett square of the cross/written account:

gametes	Bd	b^1d
b^1D	Bb^1Dd black	b^1b^1Dd cinnamon
b^1d	Bb^1dd blue	b^1b^1dd fawn

genotype with correct phenotype **[4 max]**
probability 1 black:1 cinnamon:1 blue:1 fawn **[1]**
quality of answer **[1]**

13.10 **(a)** chromosomes from a single cell/arranged in pairs (diploid cell); **[1]**
(b) 46/23 pairs; **[1]**
(c) male; **[1]**
(d) three copies/trisomy of chromosome 21; sex chromosomes XX (not XY); **[2]**
(e) **(i)** Klinefelter's: have 47/46+1 chromosomes/XXY; phenotype is male but testes are small and usually infertile/little facial hair; **[2]**
(ii) Turner's: have 45 chromosomes/one sex chromosome/XO; female, no ovaries/no secondary sexual characteristics/no breast development; **[2]**

14 Evolution

14.1 **(a)**

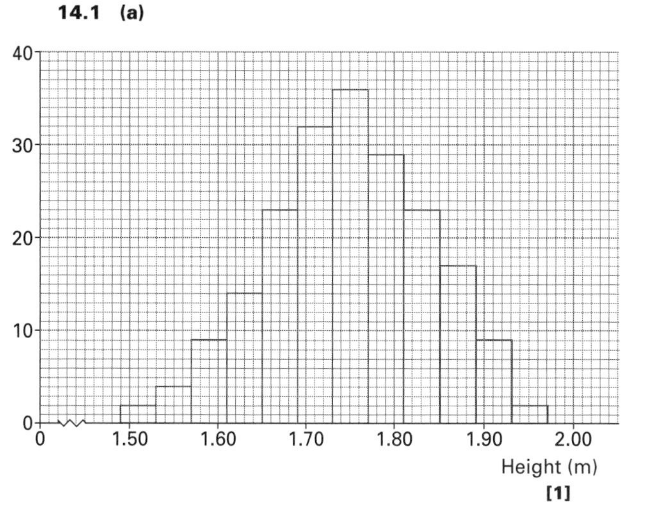

[1]

(b) **(i)** normal distribution (curve); **[1]**
(ii) the height of most individuals is in the middle of the range; few are very short or very tall/at the extremes of the range; **[2]**
(iii) mode = most common value; median = middle value; mean = average value; standard deviation = distribution of values either side of the mean/68% fall within 1 standard deviation/95% fall within 2 SD; **[4]**

(c) variation; phenotype(s); polygenes; additive; dominant; hair/skin colour, weight, shoe size, height, intelligence, sporting/musical/ dramatic ability, etc: any correct answer in any order **[max. 3]** large/significant/ sufficient; normal; distinct/separate; a single gene; discontinuously; gender/sex/tongue rolling/cystic fibrosis, etc; **[Total 14]**

14.2 **(a)** environment has little effect on characteristics determined by a single gene; discontinuous variation; example; environment more important in expression of polygenes/many alleles; show continuous variation; additive affect of dominants; example; full potential of genotype achieved only in optimum environment/vice versa; mention of studies on identical twins/separated when young. **[6]**

(b) The answers to questions like this are not always known. You are credited for making points you justify or can suggest a reasoned explanation for. For example:

A environment significant; locusts learn to behave as a group/solitary while young and exhibit this behaviour as adults;

B genetics is probably more important; appears to be inherited; left-handed parents may encourage left-handedness in their children (environment);

C environment affects expression of himalayan gene; when climate colder extremes of body develop darker hair/ and prevent heat loss; (or vice versa);

D environment significant; plants have identical genotype but it is expressed differently depending on habitat;

E environment: mirror twins/resulting from a late splitting of the zygote; (but this does not explain order of delivery); second twin is more likely to have deficiency of oxygen to brain/where right

handedness is controlled, so he learns to use left hand which is more efficient;
1 mark for each suggestion and second mark for a sensible reason **[10]**

14.3 (a) mutation = sudden change in DNA base sequence or amount of DNA in a cell;
mutant = organism carrying a mutation;
mutagen = agent that causes a mutation; **[3]**

(b) X-ray; α; β; γ rays; ultra violet radiation; chemicals/agent orange; colchicine; **[3]**

(c) (i) order of amino acids shown as:
-serine-tyrosine-lysine-alanine-glycine-tryptophan-; **[1]**

(ii) substitution; no effect/AAG and AAA both code for lysine; code is redundant/degenerate; **[2]**

(iii) –UCA-UUA-AGG-CAG-GUU-UG; **[1]**
frame shift occurs/sequence/type of amino acids change; different polypeptide formed; interfere with/stops metabolic pathway; **[2]**

(iv) UCC-AUA-UAA-GGC-AGG-UUU-G **[1]**

(d) (i) protein; **[1]**

(ii) phenylalanine would not be metabolised/would build up in cells; phenylalanine would not be converted to tyrosine; is toxic/PKU; **[2]**

(iii) albinism/being albino; **[1]**

14.4 (a) substitution; of thymine for adenine/ changing CTT to CAT; glutamine replaced by valine; resulted in slight change in β- haemoglobin; **[3]**

(b) sickle cell disease: all red blood cells sickle; severe anaemia; pain in joints/short life span; genotype homozygous for mutated allele; sickle cell trait: blood cells normal, but some will sickle at times, although heterozygous individuals have normal health on the whole; **[3 comparative points]**

(c) sickle cell trait individuals are heterozygous; alleles co-dominant; have normal and sickle cells; if infected by malaria parasite, cells sickle; sickle cells are destroyed in spleen with parasite/so person recovers; **[4]**

14.5 (a) (i) a change in the normal number of chromosomes/an extra chromosome or loss due to deletion; **[1]**

(ii) an increase in the normal number of sets of chromosomes; **[1]**

(iii) an extra copy of one chromosome; **[1]**

(b) (i) 0; **[1]**

(ii) 47/46+1; **[1]**

(iii) very rare/did not occur in sample; **[1]**

(iv) 45/46–1; **[1]**

(v) 0.005; **[1]**

(vi) 23 and 24; **[1]**

(vii) failure of homologous chromosomes to separate during metaphase 1 of meiosis; or of chromatids to separate during metaphase 2; **[2]**

(viii) meiosis 1 takes place in embryo of fetus; chromosome may become 'sticky' before meiosis is completed after fertilisation; long timespan for completion of division; **[2]**

(c) majority of *all* babies are born to younger women. **[1]**

14.6 (a) 1 = deletion; 2 = inversion; 3 = duplication; **[3]**

(b) break in Chromosome 1 separates S and T from rest; joins with Chromosome 2 which becomes longer; translocation; **[3]**

14.7 (a) (i) total number of genes in a sexually reproducing population; **[1]**

(ii) number of individuals in a population carrying a particular allele; **[1]**

(b) population is large; sexually reproducing population; random mating; population is isolated/no immigration or emigration; no natural selection; no mutations; **[5]**

(c) (i) frequency of dominant allele; **(ii)** frequency of recessive allele; **(iii)** frequency of dominant phenotype; **(iv)** frequency of recessive phenotype; **(v)** frequency of heterozygotes in population; **[5]**

(d) $p + q = 1$; **[1]**

(e) 590/6300 = 0.09365; **[1]**

(f) $q^2 = 76/139 = 0.547$; $q = 0.737$; $p + q = 1$/ $p = 0.263$;
frequency of A = 0.263, a= 0.737; **[3]**

(g) q^2 = 1 in 200/ $\frac{1}{200}$ /0.005; $q = 0.07$;
$p + q = 1$; $p = 1 - 0.07 = 0.93$;
frequency of heterozygotes = 2pq; =
$(2 \times 0.93 \times 0.07) = 13.02\%$; **[3]**

(h) $q^2 = 0.21$ / $q = 0.458$; $p = 1 - q$ / 0.542;
$p^2 = 0.293$ / frequency of MM phenotype ;
= 29.3%;
frequency of MN = 2pq / $(2 \times 0.542 \times 0.458)$ / 0.496; = 49.6%; **[3]**

(i) no – most populations do not satisfy the criteria; immigration; mutations occur; selection yes/rare; ;
[Reasons must link to your opinion. Max. 3]

(j) new variants/gene pool will increase/ natural selection; **[1]**

14.8 (a) More than one distinct phenotype exists within a population of a species at any one time. **[1]**

(b) (i) in rural woodland most moths are light morph/only 9.4% are dark morph; vice versa in industrial wood; *typica* well camouflaged against light trunks in rural area/*carbonifera* better camouflaged against industrial/sooty trunks; **[3]**

(ii) morphs least well camouflaged suffer higher predation by birds in both habitats; so moths that are well adapted/camouflaged are left to breed; in rural wood numbers of *typica*/light moth are highest; *carbonifera*/dark are lowest;or vice versa. **[4]**

(iii) air pollution has been reduced in past 50 years; would expect trees in industrial wood to be less sooty/dark; more *typica* moths could survive there; **[2]**

14.9 (a) 1 = disruptive; 2 = stabilising; 3 = directional; **[3]**

(b) stabilising selection; **[1]**

(c) **B**, **D**, **H**, **K** = stabilising selection; **C**, **G**, **J** = directional selection; **A**, **E**, **F**, **I** = disruptive selection; **[each correct 1 point]**

(d) allows mating of animals/cross breeding of plants that may not happen by natural selection; domestication of animals; increases efficiency of agriculture; breeding of animals for higher yield of meat; milk; eggs; wool; for faster maturity; bigger size; better quality of product/lean meat (Aberdeen Angus)/milk (Jersey cows); better temperament; resistant to disease;
plants that have higher yields/larger fruit; better quality crop; earlier maturity; may be cultivated more easily/shorter stems so not blown down by wind/produce crop at height that is easier to harvest; with resistance to disease; **[8]**

14.10 (a) species; hybridisation; fertile; non-viable/ not viable; sterile; donkey/ass; intraspecific; allopatric; isolated; behaviour; mechanical/ physiological; sympatric; **[12]**

(b)

	Sympatric speciation	Allopatric speciation
The most common type of speciation in animals	✗	✓
Hybrids may result	✗	✗
Occurs in populations living in the same geographical area	✓	✗
The new species may be polyploid	✓	✗
Breeding is between members of the same species	✓	✓

[5]

14.11 (a) gather fruit from higher; hands free for tasks other than walking; cooler/better protected from sun; more energy efficient; **[4]**

(b) cooking; keeping warm; live in more varied places; light when sun is down; improved technology; **[4]**

(c) dating; see sequence of changes in skeleton; to bipedalism; hands/ability to make tools; skull/size of brain; teeth and jaws/changes in diet; **[4]**

14.12

	Australopithecus	Homo habilis	Homo erectus
Tools	possibly sticks to dig up insect nests	Oldowan and pebble tools	hand axe/ cleavers/scrapers
Food collection	gatherers	gatherers and scavengers	gatherers, scavengers and hunters
Diet	vegetarian	fruits and nuts some meat	berries, fruits, roots, more meat
Use of fire	no	no	yes

[4]

14.13 (a) interbred with Cro-Magnon Man; less successful than Cro-Magnon Man at inventing new tools; at hunting; less able to adapt to changes in environment/Ice Age in N Europe; had slower birth rate than Cro-Magnons; **[3]**

(b) hands; women; **[1]**

(c) fire for light; use of natural pigments; making new pigments; tools for carving/ sculpturing; painting; detailed observation; **[5]**

(d) sanctuary; religious place; education; **[2]**

(e) teeth; bones; ivory; **[3]**

14.14 domesticated sheep/pigs/goats/cattle; had reliable source of meat/milk/leather/cloth/wool; bred in captivity so no need to hunt; crops cultivated; artificial breeding of animals and crop plants; settled communities/built homes ; support of a community; cooked; pottery developed/storage of grain; moved on after few years if site became infertile/unsuitable; **[8]**

15 Photosynthesis

15.1 **(a)** **(i)** leaves; stem; **[1]**
(ii) palisade mesophyll; spongy mesophyll; **[2]**
(b) **A** stroma; **B** double membrane/chloroplast envelope; **C** granum; **D** starch grain; **[4]**
(c) **(i)** **A**; **(ii)** **C**; **[2]**
(d) measured length = 73 mm; actual length = 73/22 000; = 0.003 318 mm/3.318 μm

15.2 **(a)** provides energy; is source of all foods; oxygen is released to replenish supplies in atmosphere/on the planet; **[3]**
(b) Water + Carbon dioxide = Glucose + Oxygen
$6H_2O + 6CO_2 = C_6H_{12}O_6 + 6O_2$ **[2]**
(c) **(i)** 1 grana/thylakoids; 2 stroma; **[2]**
(ii) solar light energy/sunlight; **[1]**
(iii) **A** water; **B** carbon dioxide; **C** oxygen; **D** glucose/starch; **[4]**
(iv) anabolic; it uses energy for synthesis of glucose; **[2]**

15.3 **(a)** red light/700nm; blue light/440nm; **[2]**
(b) they reflect green light; **[1]**
(c) absorption spectrum: obtained by passing beam of light; through pigment such as chlorophyll and then through a prism; and shows which wavelengths of light are absorbed; action spectrum: green leaves are exposed to different wavelengths of light; rate of photosynthesis is measured; most efficient for red and blue wavelengths; diagram suitably labelled for each; **[5]**
(d) pigments; separated; (paper) chromatography; filter/chromatography paper; solvent; solubility; pigment; solvent (front); Rf; **[9]**

15.4 **(a)** **A** photosystem 2/chlorophyll; **B** electron acceptor; **C** photosystem 1/chlorophyll; **D** electron acceptor; **E** reduced NADP/ $NADPH_2$; **F** oxygen; **[6]**
(b) the formation of ATP during the light dependent stage of photosynthesis; **[1]**
(c) electrons (from 'excited' chlorophyll); pass along a series of carriers; at progressively lower energy levels; energy released; is used to synthesise ATP/ADP+Pi = ATP; **[3]**
(d) hydrogen carrier/ acceptor; reduced by hydrogen from water; used in light independent stage; to reduce carbon dioxide; **[2]**
(e) **(i)** photolysis; **[1]**
(ii) water → hydrogen ions/protons + electrons + oxygen
$2H_2O \rightarrow 4H^+ + 4e^- + O_2$ **[2]**
(f) **(i)** oxygen;
(ii) ATP; reduced NADP; **[3]**

15.5 **(a)** carbon dioxide; triose phosphate/GALP; reduced NADP; ATP; **[4]**
(b) **(i)** **X** glucose; **Y** ADP + Pi; **Z** reduced NADP/$NADPH_2$;
(ii) **J** carboxylation; **K** oxidation; **L** condensation; **[3]**
(iii) carbon dioxide acceptor; **[1]**
(c) **(i)** 5; **(ii)** 2; **(iii)** 3; **(iv)** 6; **[4]**
(d) Calvin from Melvin Calvin who discovered the pathway; cycle, because the reactions continue a cyclical path; **[2]**
(e) **(i)** can accept more hydrogen; **(ii)** free to be phosphorylated to ATP; **(iii)** will be converted to biochemical compounds as required by the plant/to starch, amino acids, lipids; **[3]**

15.6 **(a)** sunlight/solar light energy; **[1]**
(b) dead/waste organic matter; **[1]**
(c) 4; **[1]**
(d) algae; green plants (which were the source of the decaying leaves); **[2]**
(e) freshwater limpet; mayfly nymph; chironomid midge larva; Hydropsyche larva; blackfly larva; freshwater shrimp; **[6]**
(f) secondary consumers; **[1]**
(g) there is insufficient energy transferred; to higher trophic levels; to support animals occupying such positions; **[2]**
(h) chain gives a single pathway; web is many food chains; shows inter-relationships/ organisms have more than one source of food; **[2]**

15.7 **(a)** **(i)** numbers of organisms; at each trophic level of food chain; **[2]**
(ii) biomass; at each level of food chain; **[2]**
(iii) energy transferred; per area per unit time; through levels of food chain; **[2]**
(b) number of organisms at that trophic level; **[1]**
(c) **(i)** X has one/few producers; Y shows large number of organisms/parasites at top level; **[2]**
(ii) X: oak tree – aphids – ladybirds – birds; **[1]** Y: grass – rabbits – mites/other parasite; **[1]** Any other correct examples.
(d) total mass of organic matter in a living organism; **[1]**
(e) **(i)** pyramid of dry mass; organisms contain different % water; **[2]**
(ii) not all organisms/complete organisms may be sampled; organisms are killed; habitat may be damaged during collection; results are true only for one point in time; **[3]**
(iii) Q is an inverted pyramid/producers are fewer than primary consumers; rate of reproduction of producers is rapid; they produce enough biomass/energy to support primary consumers; **[3]**
(f) **(i)** kJ m^{-2} yr^{-1}; **[1]**
(ii) less energy is transferred from each level to the next; **[1]**
(iii) pyramid of energy is always pyramidal shape; shows productivity/energy transfer within the chain/web over a period of time/pyramids of number and biomass relate to one point in time only; pyramids of number represent all organisms equally/and do not show biomass or productivity; **[2]**

15.8 **(a)** growth; replacement; repair of new tissues; reproduction; movement; basal metabolism; **[4]**
(b) respiration; egestion; excretion; death; **[3]**
(c) scavengers/examples; decomposers/examples; utilise the energy from waste organic materials; **[3]**
(d) reflected away from surface; unsuitable wavelength; may not strike a chlorophyll molecule; **[2]**
(e) GPP = total production of organic matter by producers; part of GPP that is available for the use of the organism/NPP = GPP – respiration/energy available for transfer to next trophic level; secondary production = new organic material produced by consumers; **[3]**
(f) **(i)** (24 950 – 2 950) = 22 000 kJ m^{-2} yr^{-1} **[1]** (you must include units for the mark)
(ii) $\frac{24\ 950}{(165\ 000 + 524\ 000 + 335\ 000)} \times 100 = 2.43\%$ **[3]**
(g) **(i)** higher temperature; greater light intensity; rain; no seasons when productivity is reduced; **[3]**
(ii) plants bred to give high yield/ productivity; crops fertilised/irrigated/ growth not limited by lack of minerals/ water; pesticides used and crop plants damaged less; **[3]**

15.9 Energy source is light from sun; absorbed by chlorophyll/photosynthetic pigments; converted to ATP; becomes incorporated into organic compounds/starch; in plants/producers; this production is gross primary production/GPP; some energy transferred from plant by respiration; net production = GPP – respiration; plant consumed by herbivore/primary consumer; herbivore consumed by secondary consumer/animal at next trophic level; organic matter transferred; some parts/bones are indigestible; at each stage energy transferred away from food web by egestion/excretion/ respiration; is used for movement/growth/ reproduction; less energy at top of chain/web; fewer animals occupy highest trophic levels; if animals or plants die/lose parts (branches/ leaves) decomposers can utilise organic matter by feeding saprophytically; or organic matter may become fossilised/if form fuel may eventually be converted to heat energy. **[10]**

16 Respiration

16.1 **(a)** A adenine; B ribose; C phosphate; **[3]**
(b) ATP + H_2O = ADP + Pi + energy/32kJ/ΔG; hydrolysis; ATPase; **[3]**
(c) stalked particles/cristae of mitochondria; **[1]**
(d) phosphorylation; endergonic; **[2]**
(e) oxidative phosphorylation (respiration); photophosphorylation (photosynthesis); **[2]**

(f)

	True	False
ATP is the universal source of energy in living organisms	✓	
ATP can be stored in living cells		✓
ATP is synthesised from ADP by a series of reactions		✓
Most ATP is synthesised by oxidative phosphorylation	✓	
The energy released when a molecule of ATP is converted to ADP is constant	✓	

(Each row must be totally correct for the mark.) **[5]**

(g) ATP is immediately available/glucose must be respired to release energy; ATP releases 32kJ/a glucose molecule may provide excess energy; **[2]**

16.2 **(a)** **(i)** removal of electrons; or hydrogen; addition of oxygen; **(ii)** addition of electrons; or hydrogen; removal of oxygen; **[6]**
(b) one reactant is oxidised and the other reduced; **[1]**
(c) reduced: **(i)**; **(iii)**; **(iv)**; oxidised: **(ii)**; **[4]**
(d) dehydrogenases; **[1]**

16.3 **(a)** cellular respiration is a metabolic pathway in cells; gaseous exchange is exchange of oxygen and carbon dioxide between the organism and the environment; **[2]**
(b) **A** catabolism; **B** exogonic; **C** metabolism; **D** anabolism; **E** endergonic; **[5]**
(c) **(i)** **W** Krebs cycle; **X** glycolysis; **Y** link reaction; **Z** electron transfer chain; **[4]**
(ii) **W** and; **Y** matrix of mitochondria; **X** cytoplasm/cytosol; **Z** cristae/stalked particles/inner mitochondrial membrane; **[4]**
(iii) mesosomes; **[1]**

16.4 **(a)** glucose; phosphorylated; ADP; three; NAD/ a hydrogen carrier; two; pyruvate/pyruvic acid; acetyl coenzyme A; link; carbon dioxide; NAD; **[11]**
(b) **(i)** pyruvate → ethanal → ethanol + carbon dioxide; **[2]**
(ii) pyruvate → lactic acid; **[2]**
(c) different cells have different energy requirements/more metabolism; more mitochondria; **[1]**
(d) during glycolysis; **[1]**
(e) muscle tissue adapts to greater energy requirement and more mitochondria are produced; **[1]**
(f) lack of energy/debility; **[1]**

16.5 **(a)** to remove carbon dioxide; **[1]**
(b) volume of oxygen being used; **[1]**
(c) control; **[1]**
(d) Q showed most reaction/greatest uptake of oxygen; mitochondria and enzymes from liver, glucose and added ATP allowed high rate of cellular respiration; P reaction slower; no added ATP; R slowest of all; only substrate available is from liver; excess ATP inhibits glycolysis; **[3]**
(e) rate of reaction doubles for 10°C rise in temperature below 40°C; **[1]**

16.6 **(a) (b)** **[10]**

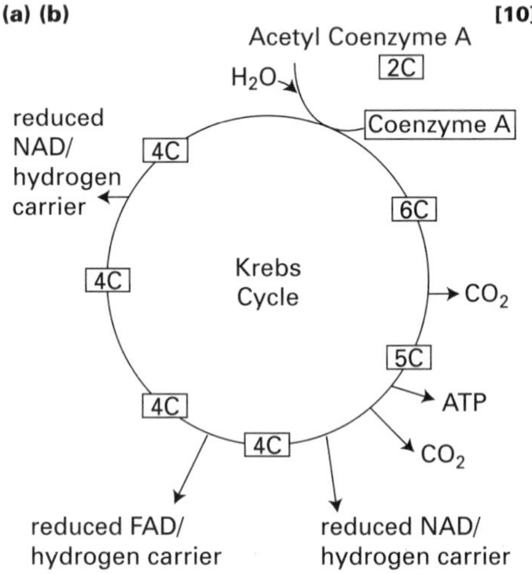

(c) two; **[1]**
(d) non-protein organic molecule; essential for functioning of an enzyme; **[1]**
(e) nicotinic acid/vitamin B complex; **[1]**
(f) transfer hydrogen to electron transport chain; can be re-used in cell; **[1]**
(g) decarboxylation; **[1]**
(h) oxidation; **[1]**

16.7 **(a)** cristae/mitochondria; **[1]**
(b) hydrogen atoms; **[1]**
(c) one molecule of ATP; **[1]**
(d) each hydrogen atom splits; into a proton/H^+ and an electron; protons pass between mitochondrial membranes; electrons continue down carrier chain; **[3]**
(e) electrons and protons join; hydrogen reduces oxygen to water; **[2]**

(f) there is a gradient; energy transferred to ATP; electrons travel from strongest reducing agent to oxygen; oxygen final acceptor; **[2]**

(g) cytochrome oxidase catalyses final step in electron carrier chain; cyanide binds with cytochrome oxidase; enzyme distorted/ unable to function; flow of electrons stops; NAD and FAD are not oxidised; ATP production stops; muscles contract/ asphyxiation/death; **[4]**

16.8 protons/H^+ actively transported; into space between inner and outer mitochondrial membranes; membranes impermeable to H^+; concentration of H^+ higher between membranes than in mitochondrial matrix; pH/ electrochemical gradient set up; at stalked particles; are hydrophilic channels; with ATPase; as pairs of H^+ pass through channels; energy is released for synthesis of a molecule of ATP/ ADP + Pi = ATP;
(Your points should be in a logical order.) **[6]**

16.9 **(a)** glycogen; **[1]**

(b) **(i)** **W** acetyl coenzyme A; **X** protein; **Y** starch; **Z** fatty acids; **[4]**

(ii) starvation/all fat reserves have been exhausted; **[1]**

(c) $RQ = \frac{\text{Volume carbon dioxide given out}}{\text{Volume of oxygen used}}$; **[1]**

(d) RQ = 114/163; = 0.69; **[2]**

(e) **(i)** they are taking in more oxygen than they are respiring; **[1]**

(ii) volume carbon dioxide given out exceeds oxygen breathed in; **[1]**

(iii) different/mixtures of substrates are respired; **[1]**

(f) no; it is not likely/possible to tell whether a pure substrate is being respired; **[2]**

16.10 **(a)** **(i)** organism capable only of anaerobic respiration; e.g. tapeworm/*Taenia*; **[2]**

(ii) organism capable of both aerobic and anaerobic respiration; e.g. yeast; **[2]**

(b) wine; beer; yoghurt; bread; cheese; silage; **[3]**

(c) **(i)** using hydrogen from reduced NAD; skeletal muscle; **[2]**

(ii) anaerobic glycolysis; **[1]**

(iii) fermentation; **[1]**

(iv) ethanol; **[1]**

(v) ATP; carbon dioxide; water; **[3]**

16.11

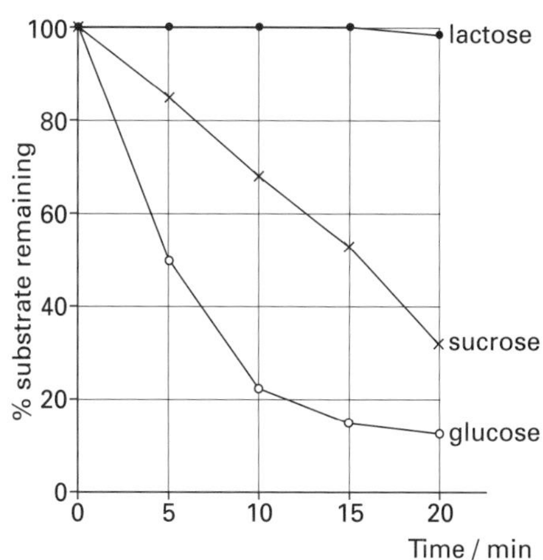

(a) axes labelled; suitable scale; correct curve for each sugar with points joined with ruler; **[4]**

(b) glucose: 30/3 = 5% substrate used min^{-1};
sucrose: 10/3 = 3.33% substrate used min^{-1};
[1 mark for each calculation and 1 for correct units. Total 3]

(c) rate slowed down and became almost stationary; substrate exhausted; **[2]**

(d) yeast does not have enzymes needed to utilise lactose; may be producing necessary enzyme at end of test; **[2]**

(e) rate of respiration would be faster; if yeast cells had synthesised suitable enzyme; gene for enzyme/lactase had been switched on; **[2]**

16.12

	Aerobic respiration	**Photosynthesis**
Which kingdoms of living organisms do this?	All	Plants; few Protoctista
Where does this occur in cells?	cytoplasm and mitochondria	chloroplasts
When does it occur?	all the time	in sunlight
How is ATP synthesised?	1 substrate phosphorylation 2 oxidative phosphorylation	photo-phosphorylation
What are the raw materials?	glucose, oxygen	carbon dioxide, water
What are the end products?	water, carbon dioxide	glucose, oxygen
Is the process anabolic or catabolic?	anabolic	catabolic

[7]

17 Ecology

17.1 **(i) B; (ii) C; (iii) E; (iv) A; (v) D;** **[5]**

17.2 **(a)** **biome** = major ecosystem of earth/ characterised by climate/vegetation; **biosphere** = parts of earth and atmosphere inhabited by living organisms; **habitat** = where an organism lives; **ecosystem** = habitats + communities living there + biotic and abiotic factors operating; **population** = all members of a species living in same habitat; **community** = all populations in a habitat; **abiotic factor** = non-living/physical and chemical/soil/climate/conditions in a habitat; **biotic factor** = due to activities of living organisms/competition in a habitat; **consumer** = feeds on organic matter made by plants directly/primary consumer or indirectly/ secondary consumer; **decomposer** = saprophyte/feeds on dead organic matter; **[10]**

(b) **ecological niche** = trophic position/feeding relationship of organism in its ecosystem; **[1]** **species** = group of organisms genetically similar; and capable of reproducing fertile offspring; **[2]** (first point only, no marks)

(c) Your answers will depend on the habitat you have chosen. The named habitat should be defined clearly e.g. mixed deciduous wood/ freshwater pond/saltmarsh, and the organisms you name must be found there! **[13]**

17.3 **(a)** progression of changes in the organisms; in a community; over time; in response to changes in environmental conditions; until a stable community/climax is achieved; example; **[4]**

(b) check these answers with a book or your tutor.

(c) final/stable community limited by climatic conditions; **[1]**

(d) **(i)** oak/deciduous woodland; **(ii)** coniferous woodland; **[2]**

(e) B primary; A secondary succession; primary succession is colonisation of a site that has not previously supported life; then succession follows; secondary succession happens on a site where organisms had lived previously/were lost; **[4]**

17.4 **(a)** sewage; fertiliser; waste from food processing factories; **[3]**

(b) organic waste releases minerals; causes algal bloom/population explosion; algae shade aquatic plants; photosynthesis decreases; algae die and are decomposed by bacteria in water; BOD/biochemical oxygen demand rises; oxygen available for organisms decreases; animals die; this process is eutrophication; **[5]**

(c) **(i)** site A = 2.46; site B = 4.11; **[2]**

(ii) B has greater species diversity; total organisms at each site similar; many tubificid worms, midge larvae at A/few at B; stonefly and mayfly larvae at B/not at A; conditions more favourable at A than B/vice versa/reference to pollution at A; **[4]**

(d) species tolerates conditions others cannot; **[1]**

(e) **(i)** aquatic habitat is polluted; **[1]**

(ii) air quality good/low concentration of sulphur dioxide; **[1]**

17.5 **(a)** **(i)** series of points lowered to ground; record number of each species touched; **[1]**

(ii) specific area within frame; organisms within it counted/cover estimated/ presence noted; **[1]**

(b) **(i)** to eliminate bias in sampling; ensure representative areas are sampled; **[2]**

(ii) allocate adjacent sides of field as axes; divide into intervals/10m; use tables/calculator to generate suitable number; of pairs of random numbers; use these as co-ordinates; sample at these points; with frame quadrat; of appropriate size; **[5]**

(iii) cover/% ground covered by each species within quadrat; frequency/% quadrats in which each species is present; abundance/use scale e.g. DAFOR (dominant, abundant, frequent, occasional, rare); **[3]**

(iv) animals move; barnacle/limpet/winkle/ earthworm/any slow moving animal; **[2]**

(c) $\frac{9}{61} = \frac{54}{\text{total snail population}} = 366$; **[2]**

(d) no snails leave the garden; no snails enter; none die/deaths = new young; marking does not increase predation on snails; **[4]**

17.6 **(a)** mark out a transect/line; use spirit level and poles to survey profile of beach; at regular intervals/5m/10m; use frame/point quadrat at each station; note species/present; **[4]**

(b) Protoctista; **[1]**

(c) desiccation/drying out; harsh temperature range; winds; waves at high tide; salt concentration; **[4]**

(d) food source; shelter when shore exposed; protection against sun; **[3]**

(e) wavelengths of visible light that penetrate deep water; are different from those on dry land; algae have different/brown pigments/little or no chlorophyll; to utilise/absorb this energy; for photosynthesis; **[3]**

17.7 **(a)** both lay more eggs early in season; Blue Tit lays high numbers early in season; this drops steeply during May; Great Tit lays fewer eggs from April until May; drop is less steep; number then stays more constant; **[3]**

(b) more food early in life; leave nest earlier and learn to survive; before current year's predators are hunting; **[1]**

(c) fledglings would be different sizes; at different stages of development; weaker/stronger; would be fewer; **[2]**

(d) plenty of food for all; will leave nest together when conditions are optimal; **[1]**

(e) less food/caterpillars have metamorphosed; birds are hungrier and noisier; greater chance of predation/more predators; **[2]**

(f) can raise brood more quickly; fewer to raise/feed; may sacrifice weaker for stronger if food short; less exhausting for parents; **[1]**

17.8 **(a)** struggle for vital resources; they all require; **[2]**

(b) interspecific: competition between different species of organism; intraspecific: competition between members of same species; **[2]**

(c) animals: food; oxygen; mates; breeding sites; **[3]**

plants: light; water; minerals; space to grow; **[4]**

(d) **(i)** density-dependent factors have a different effect depending on the density of the population/affect dense populations more severely; density-independent factors will affect the same proportion of the population/depends on the severity of the factor; **[2]**

(ii)

	Density-dependent	Density-independent
Death of cattle due to a flash flood		✓
Starvation of some members of a population due to shortage of food	✓	
Rabbits dying during an epidemic of myxomatosis		✓
Bats dying during winter where there is a shortage of suitable habitats for hibernation	✓	
Total loss of a staple crop due to drought		✓

[5]

17.9 **(a)** DDT killed many organisms besides pests; accumulated in food chain; was passed to animals at higher trophic levels; became concentrated in their tissues; caused birds to lay eggs with brittle shells/eggs never hatched; was not broken down in the environment; **[3]**

(b) kills a wide range of insects; **[1]**

(c) derived from plants; not a synthetic chemical; **[1]**

(d) pirimicab kills a specific pest; can be used with biological/environmentally friendly methods; does not kill natural predators of plant pests; but limited because it damages some crops/melons, courgettes; **[2]**

(e) managed programme which combines biological control; with limited use of chemical pesticides; selected to produce minimum ecological damage; **[2]**

17.10 Biological control: less harm to environment; slower to take effect against pest; effect not as predictable as for chemicals; needs regular application/predators die as pest numbers drop/cycle of predator–pest; may need to manage local area to encourage natural predators; examples;

Chemical control: effect on pests more predictable; single/few applications necessary; takes effect quickly; can persist in environment; may enter food web and so affect animals at higher trophic levels including humans; may kill organisms other than intended pest; some are selective; including natural predators; pests may become resistant to chemical; use examples;

(Your answer should be logical and well balanced to gain the full marks.) **[10]**

17.11 **(a)** constituent of amino acids/proteins; **[1]**

(b) 80%; inert/unreactive; due to triple bond between two nitrogen atoms; **[3]**

(c) **V** *Azotobacter/Clostridium/Bacillus*; **W** *Nitrobacter*; **X** *Nitrosomonas*; **[3]**

(d) an organism that synthesises its own organic food from simple inorganic materials; and derives the necessary energy from chemical reactions; *Nitrobacter/Nitrosomonas*; **[3]**

(e) Decomposition; of plant and animal protein remains/waste products; by fungi/detritivores; producing ammonium compounds; **[3]**

(f) Relationship is mutualism/is nutritional benefit to both organisms; plant has source of ammonium ions from *Rhizobium*; bacterium obtains carbohydrate from plant; **[3]**

(g) *Thiobacillus/Pseudomonas denitrificans*; anaerobic soil/lack of oxygen; wet/waterlogged soil; cold; **[3]**

(h) during a flash of lightning; nitrogen forms oxides; which react with water in atmosphere; form nitrogen acids; wash into soil; **[2]**

(i) addition of fertilizers/named fertilizer/manure; **[1]**

18 Humans and environment

18.1 **(a)** the population is greatest (around 15 million people); and the rate of growth is the largest; **[2]**

(b) more people live independently (younger people want to be house owners, single parents, single elderly people and fewer extended families); increasing demand for homes; **[2]**

(c) first North West, then West Midlands and North East; rural, less employment, fewer transport links; draw of economic expansion in South East; companies tend to set up main offices in South East; loss of traditional manufacturing base; e.g. steel, etc **[5]**

(d) traffic congestion/air quality reduces; more road building/loss of land; more house building/loss of land; industrialisation/waste disposal increases; more commercial activity e.g. retail parks/loss of land and increase in traffic; etc **[4]**

18.2 **(a)** biodiversity describes the number of different species found in any particular ecosystem; relates to total gene pool of an ecosystem; offering potential for variation; might provide useful genes in the future; biomes include hot desert; – having numerous invertebrates, reptiles such as snakes and some small mammals such as desert rat, which can cope with drought conditions; savannah; – tropical grasslands which provide grazing for large mammals such as elephants and zebra as well as invertebrates such as termites; temperate forest; – variety of birds and small mammals such as voles, squirrels etc, many invertebrates such as earthworms; tropical rain forest; – huge number of invertebrate and vertebrate species such as primates/huge biodiversity; tundra; – cold conditions e.g. arctic; dry because water becomes ice, mammals such as arctic fox and migrating birds in summer;

[any 10 relevant points fluently linked, including other biomes]

(b) biodiversity represents a gene pool; genes code for different characteristics which might be useful in the future; e.g. genes that offer resistance to disease; genes which produce medicinal substances; genes which produce useful materials etc; **[2]**

(c) tends to reduce biodiversity; **[1]**

(d) large areas of land turned into monoculture; with a dominant species such as wheat or other arable crop; which influences which species can coexist; hedgerows and patches of scrub and woodland removed; pesticides kill weed species; **[2]**

18.3 **(a)** to protect natural environment; increase sustainability of forestry; gain credibility internationally; or other relevant; **[2]**

(b) reduces biodiversity/number of species in that environment; **[1]**

(c) much variation within small areas of rainforest/many species only represented by a few individuals; so removing their environment ends the existence of those species; **[2]**

(d) only mature trees removed so younger ones remain to grow to full height; other species existing on younger individuals of the logged species do not lose their environment completely; less disruption of natural environment; soil erosion minimised; climatic changes minimised; **[3]**

(e) more carbon dioxide remains in atmosphere; less oxygen put into atmosphere; more moisture fluctuation/lower humidity **[2]**

(f) less roots to hold soil structure; so rain washes loose topsoil into waterways; **[2]**

(g) international funding to provide education; financial support for people starting alternative trade; international task force to observe and monitor what is going on; exchange of scientific knowledge, etc; **[2]**

18.4 **(a)** protects natural environment from industrial development and house building; and other changes brought about by inappropriate use of land; and helps development of land use for leisure amenity in appropriate manner; **[2]**

(b) protects particular species; which may nest, feed or form a vital part of a food chain at particular site; and prevents environmental alteration; such as drainage of a marsh, or inappropriate use of land such as industrial or residential building; **[2]**

(c) provides shelter, food and nesting space; and tends to increase biodiversity; **[2]**

(d) ensures suitable water level; for bird species which feed such as wading birds; or nest amongst reeds; and for invertebrates in the estuarine mud; and might protect sand dunes from erosion; **[2]**

(e) helps guide walkers; increases amenity use; without damage to sensitive areas; less disturbance to sensitive species such as nesting birds; or similar answer; **[2]**

(f) helps protect reproductive fish stock; so future supplies are ensured; prevents over fishing; **[2]**

(g) managing forests; so that some trees are removed on a rotation basis; but whole areas are not removed at one time; and using space in between large tree species for other cropping; living and using the forest without destroying it; **[2]**

(h) bracken spreads over large areas; and kills other species, reducing biodiversity; **[2]**

(i) reduces deer population; reduces damage to trees; means more food supplies for deer which are not culled; and greater health of deer population; **[2]**

(j) stops dune erosion; stabilises dune flora and fauna; conserves coastal ecosystems; **[2]**

18.5 **(a)** **industry**: heat e.g. hot water discharges; particulates such as paper fibres; chemicals such as organic solvents or acidic solutions; **[2]**

farming: organic materials such as pig manure, slurry or run off from manure heaps; chemical residues such as pesticides or fertilisers **[2]**

domestic: garbage such as litter which blows into waterway; detergent from discharged treated water; **[2]**

burning fossil fuels: acid non-metal oxides such as sulphur dioxide; and oxides of nitrogen; which dissolve in rain water; **[1]**

(b) species act as bio-indicators; e.g. stonefly requires unpolluted water with good oxygen levels; tubificid worms can live in polluted waters that lack oxygen; **[2]**

18.6 **(a)** rapid growth of algae; related to high mineral content of water; floats on water surface **[2]**

(b) death of algal bloom; causes oxygen depletion as it decays; **[2]**

(c) lowering of pH of waterway or soil; often caused by soluble non-metal oxides; such as sulphur dioxide; or deposits of plant materials such as pine needles; **[2]**

(d) oxygen demand caused by decay of organic materials; measured in mg dm^{-1}; **[2]**

(e) species which are senstitive to a specific level of pollution; will not appear in an ecosystem if it is polluted beyond a certain level; acts as a quick reference scale; shows effect of pollution on biodiversity; **[2]**

18.7 **(a)** **(i)** planet becomes warmer at Earth's surface due to atmospheric changes; **[1]**

(ii) gas which prevents the radiation of infra red waves back into outer space; **[1]**

(b) there is much more of it; **[1]**

(c) decaying organic material e.g. landfill sites; decaying materials in waterlogged situations e.g. rice fields; etc **[1]**

(d) in refrigerator coolant systems/as propellants in aerosol cans; **[1]**

(e) rise in sea level; polar ice caps melt; shift of climatic zones so Northern hemisphere becomes warmer; drought and storms/floods in areas which do not usually experience them; **[3]**

(f) atmosphere is shared globally; human activity in one country affects people in other countries; greater effectiveness of international action; **[2]**

18.8 oxidation; coal; gas; oil; oxides; sulphur dioxide; nitrogen oxides (nitrous oxide, nitric oxide, nitrogen dioxide); acidic; decay/decomposition/ rotting; smaller/lesser/ reduced; photosynthesis; **[11]**

18.9 **(a)** carbon dioxide; **[1]** **(b)** nitrogen dioxide; **[1]**
(c) nitric oxide; **[1]** **(d)** carbon monoxide; **[1]**
(e) sulphur dioxide; **[1]** **(f)** nitrous oxide; **[1]**

18.10 **(a)** increase in levels of chemicals such as PCBs in atmosphere; above two poles/Arctic and Antarctic/Australia; **[2]**

(b) more UV rays reach the Earth's surface; harmful to skin/retina/causes more sunburn/skin cancers; **[2]**

(c) health hazard/greenhouse gas; **[2]**

18.11 find renewable energy sources/alternatives to fossil fuel sources; improve public transport and subsidise cost; control road building; increase fuel prices; increase use of rail links for freight; educate/promote environmental awareness; car-sharing schemes; encourage employers to give incentives for using public transport to city locations; etc. **[8]**

18.12 **(a)** rises during February and falls after August; most mites appear in July; varies from less than 5 to 25 mites per 5 cm^2; peaks and troughs are large; **[3]**

(b) points where pesticide is applied; causes rapid death of most mites; **[2]**

(c) using one species which is a natural predator or parasite of another to control the occurrence of the second species; **[1]**

(d) more steady; lower than chemically controlled plot; higher at start; declines from June; **[2]**

(e) **(i)** cheaper; better environmentally as no pesticide residues will occur; marketing advantage of organic produce; safer for workers; **[3]**

(ii) faster; have more belief in it; no need to use living control agents; easier to time; **[3]**

18.13 **(a)** larger catches due to more effective techniques; **[1]**

(b) over-fishing which reduces overall stock size; and less fish reaching maturity affecting reproductive stocks; **[2]**

(c) affects people's livelihood/earnings which influences employment figures influences voting at general election; **[1]**

(d) smaller catches; travelling further for better fishing grounds; required to monitor catches more closely; return greater proportion of fish to sea; expense of different equipment; etc **[2]**

(e) allow more fish to reach reproductive maturity; allow natural stocks to build up; be more stable economically; **[3]**

(f) restricting catches to a particular time of year; restricting catches to mature fish; rest periods from fishing; fishing different species; etc **[1]**

19 Reproduction

19.1 **(a)** gamete producing organs: testes and ovaries; **[1]**

(b) glands and tissues such as seminal vesicles, breasts, uterus, etc; **[1]**

(c) haploid sex cell; **[1]**

(d) body tissue cell; **[1]**

(e) contains half genetic complement/ chromosomes single not in pairs; **[1]**

(f) diploid cell resulting from fertilisation of gamete; **[1]**

19.2 **(a)** produce offspring to continue the species/ increase population; allows genetic variation and possibility of stronger characteristics; **[2]**

(b) family size is decreasing; greatest % change is Latin America/Caribbean; in industrialised countries birth rate is decreasing faster than in developing semi-industrialised countries; birth rate is lower in industrialised countries than parts of Africa, Asia or Caribbean **[3]**

(c) availability and awareness/use of contraception; fertility, health, economic situation, political factors such as legislation or war; religion/culture; **[4]**

(d) environmental factors affecting survival of individuals; in some countries support is provided for those who would otherwise not survive, informally via community or formally by government intervention; population increasing as survival rate increases; **[3]**

19.3 **(a)** **A** testis – sperm production/gametogenesis **[2]**

B urethra – muscular duct carries/propels sperm along penis/out of body **[2]**

C scrotum – skin folds enclosing testes externally **[2]**

D seminal vesicle – secretes fluid containing various chemicals (which add to sperm viability) **[2]**

E vas deferens – muscular tubes carry/propel sperm from epididymis to urethra **[2]**

F epididymis – temporary store for sperms **[2]**

(b)

Feature	Built of diploid tissue	Produce haploid cells	Implantation occurs here	Fertilisation occurs here	Lost during men-struation
Endometrium	✓	✗	✓	✗	✓
Cervix	✓	✗	✗	✗	✗
Ovary	✓	✓	✗	✗	✗
Vagina	✓	✗	✗	✗	✗
Oviduct	✓	✗	✗	✓	✗
Fimbriae	✓	✗	✗	✗	✗

(c) Cells from the uterus are grown/cultured; and examined microscopically for abnormality; can detect cell changes in pre-cancerous and cancerous cells; so help early diagnosis; **[3]**

19.4 **(a)** gametes/sex cells; ova; germinal layer; mitosis; oogonia; diploid; primary oocyte; puberty; secondary oocytes; haploid; fertilisation; **[11]**

(b) **A** ovum/egg; **B** secondary oocyte; **C** primary follicle; **D** degenerating corpus luteum; **E** developing corpus luteum; **F** germinal layer; **[6]**

(c) 1 = 2n; 2 = 2n; 3 = 2n; 4 = 2n; 5 = n; 6 = n; **[6]**

19.5 **A** sperm (atozoa); **B** spermatocytes; **C** spermatogonia; **D** germinal layer; **E** position of nutritive cell; **[5]**

19.6 **(i)** methods relying on preventing contact between gametes; includes natural rhythm/ no intercourse during fertile period; withdrawing penis before ejaculation; condom; diaphragm or cap; as physical barriers; also sterilisation which prevents male gametes travelling normal route; **(ii)** chemical prevention; includes spermicide creams which kill sperms; and the pill which may prevent ovulation; **(iii)** IUD; prevents implantation by making interior of uterus unsuitable;

[any 10 points fluently put together]

19.7 **(a)** acrosome and nucleus in the head; middle part containing mitochondria; tail; **[3]**

(b) acrosome ruptures to release enzymes; which digest area of zona pellucida allowing sperm to enter; **[2]**

(c) change in electrical properties of surface membrane; thickening of zona pellucida; **[2]**

(d) meiosis is completed/second meiotic division takes place; **[1]**

(e) mitosis; **[1]**

(f) morula; **[1]**

(g) implantation; **[1]**

(h) genetic abnormality of embryo; congenital defect of fetus; uterine abnormality e.g. fibroids; maternal physiology affects embryo adversely; **[3]**

19.8 **(a)** identical twins are formed from fertilisation of one oocyte/one original zygote; which splits at blastocyst stage; non-identical twins involve fertilisation of two oocytes; giving two zygotes; **[4]**

(b) umbilical cords may become entwined/ tangled; and affect blood supply to embryos; **[2]**

(c) formation of twins from two separate zygotes; **[1]**

19.9 cystic fibrosis is an inherited disease; a single allele can be carried without symptoms; but inheriting two causes the disease; both parents could be genetically screened; to look for the cystic fibrosis gene; genetic counselling is information and support; helping people to decide whether to have a family or not; to understand the probability of having a child with cystic fibrosis; whether to have the fetus genetically screened before birth; the options if an unborn child does have cystic fibrosis; what the procedures are for coping with a child who has cystic fibrosis; **[5]**

19.10 **(a)** **nutrition**: food materials diffuse from maternal blood to fetal blood; include glucose, amino acids, vitamins and minerals; **respiration**: oxygen diffuses from maternal to fetal blood;

excretion: carbon dioxide, urea and other wastes diffuse out into maternal blood; **protection**: against entry of some bacteria and viruses; antibodies provide immunity; produces progesterone, prostoglandins and oestrogen. **[10]**

(b) outgrowths of chorion membrane; into endometrium/lining of womb; increase surface area for diffusion of materials; **[2]**

(c) molecules at surface; **[1]**

(d) endothelium of fetal capillary wall; **[1]**

(e) drugs; some viruses such as rubella; **[2]**

19.11 (a) First stage: dilation of cervix to around 10 cm diameter **[2]** 1 = ACTH **[1]**
2 = prostaglandins **[1]**
Second stage: contraction uterine wall/abdominal muscles pushes baby out along vagina **[2]** 3 = stretch receptors **[1]**
4 = oxytocin **[1]**
Third stage: placenta detaches and is pushed out of uterus **[2]**

(b) makes muscle fibres in epithelial tissue contract and help push out milk **[2]**

(c) prolactin; stimulates lactation; **[2]**

19.12 (a)

	Breast milk	Formula milk
Nutrition	provides all nutritional needs in exact proportions required	provides all nutritional needs in similar proportions to human milk could be made up wrongly
Convenience	baby can be fed at any time only mother can feed baby	need to have feeds prepared or be able to make/warm/them up anyone can feed baby
Hygiene considerations	need for basic hygiene	need for careful hygiene/sterilisation, clean water source
Immunity	immunoglobulins in colostrum protect against infection	cannot provide immunity
Cost	free	has to be purchased

[10 relevant points]

(b) Highest in US, UK and Netherlands; lowest in Poland and Sweden; not necessarily related to wealth; more likely to be cultural; or to do with lifestyle factors such as two parents working; even highest rate shown is less than 50% which reflects preference of parents for breast feeding/fewer people bottle feed than breast feed; **[5]**

19.13 (a) all tissues have increased rate of respiration because of demands of fetus; e.g. kidney has to filter additional waste produced by fetus; heart has to deliver more nutrients/oxygen and remove more waste; through placenta which contains large blood supply; uterine muscle is enlarged and thickened; **[4]**

(b) 21.6 $cm^3 min^{-1}$ (include unit) **[2]**

(c) additional demand for oxygen and increased circulation; raises mother's cardiac output; **[2]**

(d) higher metabolic rate; **[1]**

20 Life stages

20.1 (a) helps normal neural tube and spine development; and protects against neural tube defects such as spina bifida; **[2]**

(b) microgram 0.000 001 kg or 0.001 g **[1]**

(c) brussels sprouts, green beans, oranges, yeast, beef extracts; **[1]**

(d) breakfast cereal/bread; **[1]**

20.2 (a) permanent increase in size; **[1]**

(b) difference in shape because graph A shows the total change in mass over time/a rate; while graph B shows fluctuation in the rate of change of mass over time/a change in rate; **[2]**

(c) cell division produces more cells; which increase in size/cell expansion occurs; as contents accumulate; **[3]**

(d) blood/skin grow throughout life; bone grows most rapidly in childhood; **[2]**

(e) supine length/how tall an infant is; skull circumference; body mass; **[3]**

(f) body size increases throughout childhood and into adolescence; head is larger in proportion to rest of body in infancy; limbs similar proportion to main body/trunk; abdomen large in infancy as liver grows faster in infancy than rest of life; **[4]**

20.3 (a) hypothalamus; pituitary; **[2]**

(b) proteins needed for membrane production; allowing cell division/cell growth; **[2]**

(c) GH now readily available/mass produced; as it is harvested from/produced by genetically engineered bacteria; **[2]**

(d) TSH affects thyroid; stimulating production of thyroxine; **[2]**

(e) ovary; after puberty; **[2]**

(f) stimulates metabolism of protein and carbohydrate; so more resources available for growth; **[2]**

20.4 **(a)** steepest slope equates to fastest rate; **[1]**

(b) up to one year; **[1]**

(c) increases after 11 years; then slows at 13 years; **[2]**

(d) show similar curve up to 10 years; lower line for years 10 to 14; then higher line after 14 years; **[3]**

(e) similar rate of growth, boys slightly heavier (but shorter) than girls up to 10 years old; girls have growth spurt at 11; so overtake boys; boys have growth spurt at 14; and overtake girls **[4]**

20.5 adolescence; threshold/certain level; primary; gametes/sex cells; hormones; testosterone; females; characteristics; breasts/widening hips; muscle; **[10]**

20.6 **(a)** FSH/follicle stimulating hormone; **[1]**

(b) depresses/inhibits the action of; **[1]**

(c) once a follicle is maturing, there is no need for others to mature at the same time; **[1]**

(d) produces progesterone/maintains endometrium/keeps lining of uterus thick; **[1]**

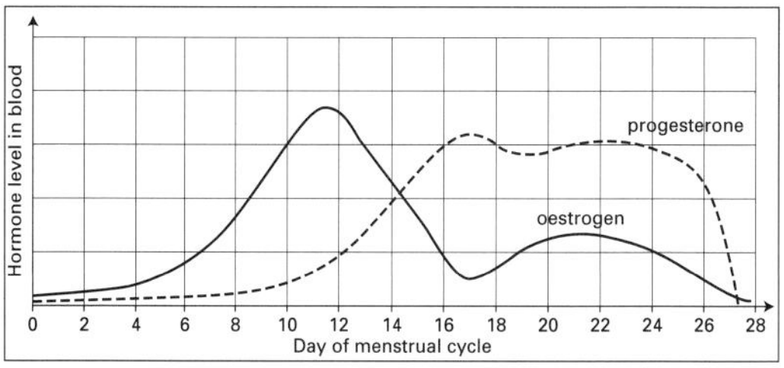

20.7 **(a)** line is at a lower level than oestrogen up to day 14; and then rises and stays at the higher level to around day 27; **[2]**

(b) corpus luteum; **[1]**

(c) corpus luteum shrinks; no need for thickened lining if fertilisation has not occurred; **[2]**

(d) fertilisation has occurred; **[1]**

20.8 ageing caused by decline in rate of cell division; slower recovery rate from infections or injury; osteoporosis may occur; loss of bone tissue/lowered bone density; makes bones brittle and more likely to fracture; muscles lose tone; so not as strong; collagen reduced in skin tissue; so wrinkles develop; cardiac output lowered; as metabolic rate slows; menopause in females; menstruation stops; prostate enlargement in males; **[10]**

20.9 **(a)** **osteoporosis** is loss of bone density; due to mineral/calcium loss; and protein matrix loss; **osteoarthritis** is an inflammation of the joints; affecting cartilage (which breaks down); ligaments (which become rigid); and bone surfaces (which become eroded); **[4]**

(b) neurodegenerative disease acute: stroke chonic: Alzheimer's disease; various dementias, Parkinson's disease and multiple sclerosis (MS) can be either acute or chronic in same person at different times.

20.10 **(a)** 15 per 1000 population; **[1]**

(b) more heart disease; greater proportion of smokers; larger body mass places a greater strain on the heart; **[2]**

(c) the mortality rate is high at birth; then decreases to the age of one year; although it is still higher than the rest of childhood; from one year the rate remains static up to about 15 years; **[4]**

(d) infant is more vulnerable during a period of illness because of small body mass; and has not developed as much immunity to infection; **[2]**

(e) mortality rate rises; and starts to diverge – rate increases faster in males than in females; **[2]**

(f) people live longer; so both curves would be lower; and curve might be flatter; **[2]**

(g) **(i)** represents quality of life; reflects how long people can be independent; **[2]**

(ii) disease; poverty; war; lack of medical intervention; etc **[2]**

(iii) highly processed diet; obesity; heart disease; **[2]**

21 Digestion and nutrition

21.1 **(a)** autotroph synthesises organic food from simple inorganic materials and energy; heteroptrophs feed on other organisms; holozoic nutrition is ingesting solid organic matter from other organisms, digestion and absorption of soluble products; **[3]**

(b) ingestion = taking in of food; digestion = physical and chemical breakdown/hydrolysis of large molecules to small soluble molecules; egestion = expulsion of undigested food as faeces; **[3]**

(c)

Site	Active enzyme	Substrate	Product(s)	pH
Mouth	salivary amylase	starch;	maltose;	7.0;
Stomach	1. pepsin 2. rennin;	protein; caseinogen	polypeptides; casein;	1.0– 2.0;
Duodenum	1. trypsin; 2. pancreatic amylase; 3. pancreatic lipase	polypeptides; starch; lipids;	peptides maltose fatty acids and glycerol;	7.0– 8.0;
Ileum	1. lactase; 2. sucrase; 3. maltase	lactose sucrose; maltose;	glucose, galactose; glucose fructose glucose;	7.0– 8.0;

[22]

(d) (i) **X** exopeptidase; **Y** endopeptidase; **Z** peptide bond; **[3]**

(ii) exopeptidases hydrolyse the terminal bonds of amino acids at the ends of the chain; endopeptidases split bonds between amino acids within the chain; by hydrolysis; trypsin; **[3]**

(e) extracellular digestion: in lumen of ileum; by enzymes in pancreatic juice; intracellular digestion: of smaller/partly digested molecules within epithelial cells; in vacuoles; **[2]**

21.2 **(a)** folded surface of mucosa; **[1]**

(b) secretion of digestive juices/enzymes; **[1]**

(c) contract and relax; to move food along/mix food and digestive juices; peristalsis; **[3]**

(d) serosa/connective tissue; **[1]**

21.3 **(a)** ileum; **[1]**

(b) large surface area; thin epithelium; rich capillary network; **[3]**

(c) peristalsis brings more digested food; rich blood supply transports away glucose and amino acids; active transport into epithelial cells; **[3]**

(d) (i) pass into epithelium lining; by facilitated diffusion; to blood capillaries; sodium potassium pump helps maintain concentration gradient; **[3]**

(ii) form micelles; dissolve in cell membranes; form triglycerides and phospholipids; enter lacteals; lead to lymph system; then to venous system; **[3]**

21.4 **(a)** lactose not digested; to glucose and galactose; remains in gut; osmotic gradient not established; water not absorbed; **[2]**

(b) reduced surface area; malabsorption of nutrients and salt; remain in gut; water not absorbed; **[2]**

(c) elderly; **[1]**

(d) water passed through gut too quickly for absorption; unable to pass into tissues by osmosis; **[1]**

(e) expertise/trained medical staff needed; specialised equipment; cost; mostly available in developed countries; **[2]**

(f) cheap; can be made easily; no specialised staff/equipment necessary; **[2]**

(g) add starch; breaks down to glucose more slowly; stays longer in gut; so allows time for more water to be absorbed; **[2]**

(h) solution in gut has higher osmotic potential than body tissues; which lose water by osmosis; body becomes dehydrated; **[2]**

21.5 **(a)** reflex action; **(b)** hormone; pepsinogen; hydrochloric acid; **(c)** secretin; CCKPZ; **(d)** bile; secretin; **(e)** CCKPZ; **(f)** CCKPZ; secretin; **(g)** enterogastrone; **[12]**

21.6 **(a)** varied food providing correct proportions of nutrients; energy content; for age of individual; their activity; appropriate to their health; mostly carbohydrate; fruit and vegetables; some protein/dairy products; limited fat/sugar/salt; **[5]**

(b) cellulose; hemicellulose; pectin; **[2]**

(c) laxative; reduces blood cholesterol; slow release of sugar from starch reduces rapid rise in blood sugar after eating; fermented by bacteria in colon to add to energy; protects against bowel cancer; **[3]**

(d) cultural/traditional; religious; moral; dislike/allergic to animal products; **[2]**

(e) no natural source of vitamin B_{12}; very low levels of calcium in vegan foods; **[2]**

(f) BMR is lower; have less muscle/lean tissue/more fat; exercise less/less energetically; **[2]**

(g) Males usually higher body mass; have more muscle/less fat; have higher BMR; or vice versa **[2]**

(h) advice government gives to public is standardised; health professionals use same standards; used in planning diets; for food labelling; in surveys; **[3]**

21.7 **(a)** minimum energy required; to maintain metabolism; while body is at rest; **[2]**

(b) insulin; thyroxine; adrenalin; noradrenalin;

(c) **[2]**

	True	False
Women have a faster resting BMR than men		✓
BMR is lower in older people	✓	
BMR is higher in people living in a cold climate than in a temperate climate	✓	
Following a serious injury BMR generally falls		✓
Metabolic rate rises after eating a meal	✓	

[5]

(d) **(i)** energy requirement of fetus; and for changes in mother's body; **[2]**

(ii) exercise increases proportion of muscle/lean tissue in body; muscle has high metabolic activity; even when not exercising; **[1]**

(iii) obese women had more cells; need more energy to metabolise; use more energy in movement; **[2]**

21.8 folate/folic acid; green vegetables; cereals; three; fats; proteins; placenta/breasts; blood; absorption; green vegetables/dried fruit/nuts; anaemic/iron deficient; menstruation; C; citrus fruit; 700; D; sun; 800 kJ; protein/liquid; milk secretion; iron; 850; **[22]**

21.9 **(a)** marginally overweight; **[1]**

(b) **(i)** $80 \times 121/100$ or $80 \times 1.21 = 96.8$ kg; **[3]**

(ii) $96.8/3.168 = 30.5$ **[3]**

(c) 20–25; **[1]**

(d) it is an objective measurement; used widely by health professionals; **[1]**

(e) diabetes; heart disease; arthritis; breathing problems; gout; **[4]**

(f) **(i)** they develop more muscle which has more mass than fat/adipose tissue; **[1]**

(ii) more lean tissue; less fat; more skeletal tissue; **[3]**

(g) anorexia nervosa; **[1]**

21.10 BMI below 16; more common in females/young women; intake of calories is less than energy expended; condition is addictive; causes periods to stop/amenorrhoea; infertility; low blood sugar; imbalance of minerals/lack of calcium causes decrease in bone density; feel cold; depression and anxiety; have low self esteem; sufferers believe they look fatter than they are/psychological disorder; behaviour is secretive; Bulimia person often has normal BMI; secretly binges; then vomits in secret; may abuse laxatives/diuretics; **[10]**

22 Homeostasis: principles and examples

22.1 **(a)** **(i)** conditions surrounding cell/bathing fluid such as tissue fluid/outside the cell membrane; **[1]**

(ii) conditions within cell, such as pH, O_2 tension, etc; **[1]**

(b) temperature; affects enzyme systems/metabolic pathways; water potential/salt/ion/glucose concentration; affects movement of water in and out of cell; **[4]**

(c) protein molecules altered by inappropriate pH; affects enzyme systems/metabolic pathways; **[2]**

22.2 **(a)** conditions change within the body; as a condition (reference to specific example) e.g. level of glucose in blood; moves away from optimum; mechanism kicks in to stop production of more of that condition; negative feedback stops a condition moving further from optimum; moves past optimum in other direction; too low not too high; mechanism pulls level back to optimum again; **[4]**

(b) receptors on surface membrane; **[1]**

(c) brings about a change; **[1]**

22.3

	Main organ systems involved in homeostasis					
Statements	**Kidney**	**Hypothalamus**	**Skin**	**Liver**	**Pituitary**	**Thyroid**
Acts as a master gland	✗	✗	✗	✗	✓	✗
Regulates water, salt and pH in body fluids	✓	✗	✗	✗	✗	✗
Regulates body temperature	✗	✗	✓	✗	✗	✗
Controls the pituitary gland	✗	✓	✗	✗	✗	✗
Regulates the metabolic rate	✗	✗	✗	✗	✗	✓
Regulates carbohydrate metabolism	✗	✗	✗	✓	✗	✗

(whole line correct for one point) **[6]**

22.4 hyperglycaemia: glucose level in blood higher than optimum/too high **[2]** and hypoglycaemia: glucose level in blood too low/lower than optimum **[2]**

22.5

Effects on	Insulin	Glucagon
rate of cellular respiration	increases	none
conversion glycogen to glucose	none	increases
conversion glucose to glycogen	increases	none
rate of uptake of glucose by cells	increases	none
rate of conversion of glucose to fat	increases	none
rate of conversion of amino acids to glucose	none	increases

[any 8 correct points]

22.6 liver has vital role as regulator of blood glucose level; produces insulin and glucagon; mention interconversion of glucose and glycogen; link to fat metabolism; also produces bile/role in emulsifying fats and fat digestion; breaks down excess amino acids/deamination; and produces urea; role in protein metabolism via transamination; and production proteins such as clotting factors; storage function e.g. minerals, vitamins; and blood reservoir; breaks down hormones; detoxifies alcohol, nicotine, etc; heating effect because of rapid metabolism; **[10]**

22.7 **(a)** damage to liver cells; causing liver disease which can reverse if alcohol consumption is lowered; cirrhosis; of the liver as liver cells are replaced with connective tissue; sometimes liver cancer; **[3]**

(b) caffeine; paracetamol; codeine; steroids (illegal and medical drugs); nicotine; **[2]**

(c) following injury or surgery; the liver can regenerate additional tissue; **[2]**

(d) under normal circumstances the liver breaks down bile; **[1]** including the pigments it contains such as bilirubin; build up of bile pigments causes yellow coloration and jaundice; **[1]**

(e) hepatic artery to liver; hepatic vein from liver to vena cava; hepatic portal vein from gut (spleen, stomach, pancreas, small intestines, colon and rectum) to liver; **[2]**

22.8 (Refer to standard text)

22.9 ion/salt; excretion; pH; filtration/ultrafiltration; glucose; molecular size/relative molecular mass/ RMM; reabsorption; mitochondria; osmosis; vasa recta; active transport; water potential; distal/second convolluted tubules; ADH/ antidiuretic hormone; **[14]**

22.10 **(a)** 180 dm^3 less the amounts shown in the table = 1.0 dm^3

[1 for correct calculation and 1 for unit]

(b) reduced volume of urine; as additional loss by sweating is likely; **[2]**

(c) salt absorbed via blood into body tissues, lowering water potential; water passes by osmosis into blood and body cells along osmotic gradient/water follows salts into tissues; requires additional water to reduce water potential in blood and tissues; **[3]**

22.11 **(a)** diffuses into dialysate/dialysis solution while passing through dialyser/kidney machine; because urea concentration in blood is higher than in dialysate/following urea diffusion gradient; **[2]**

(b) glucose concentration in blood and dialysate are equal; so diffusion in either direction is equal; **[2]**

(c) plasma proteins cannot pass through membrane; so does not alter; **[2]**

(d) counter flow is efficient; because it means that as blood progresses through dialyser it meets cleaner and cleaner dialysate; maintains diffusion gradient throughout machine; **[2]**

(e) removes gas/air bubbles; removes clots; **[2]**

22.12 **(a)** gains and losses by: convection; conduction; radiation; also heat input from respiration; heat loss due to sweating; **[5]**

(b) brain and organs within thorax and abdomen constitute body core which maintains an even temperature to protect vital organs; **[1]**

(c) 36.8 (degrees)/37(degrees) C; **[1]**

(d) illness/infection; temperature extremes that the body's mechanisms cannot cope with; **[2]**

(e) larger surface area to volume ratio; more of surface is near the surroundings; so temperature gradient has greater effect; **[2]**

(f) fluctuation over a 24h period/day and night; **[1]**

(g) middle of the night 2–4am; deep sleep; thermoregulatory mechanisms depressed; less active; **[3]**

(first point and any other two)

22.13 **(a)** melanin is a protective skin pigment/ produced by melanocytes in skin; in response to exposure to sun; **[2]**
photoageing is accelerated skin ageing caused by solar damage to basal skin cells; collagen loss from deeper skin layers and increased melanin pigmentation; **[2]**
melanoma is a type of skin cancer; enhanced risk results from solar damage; **[2]**
scar tissue develops following injury, contains additional connective tissue so appears lumpy; does not contain melanocytes; **[2]**
erector pili muscles occur at base of hair follicle; cause hair erection; **[2]**
sebaceous gland occurs towards base of hair shaft; produces sebum which oils and conditions skin; **[2]**

(b) sketch shows connecting/shunt vein between the small arteries in the lower skin layers and below to be constricted; and capillary network receives more blood (show as broader); and more heat is radiated (show arrows larger); **[3]**

(c) body temperature falls below 35°C; and hypothalamus stops functioning as thermostat; so temperature falls lower and lower; babies; and elderly; are vulnerable as thermoregulation not as efficient **[5]**

(d) heat gain centre detects drop in temperature and stimulates responses in skin; such as vasoconstriction and reduction in sweating; so less heat is transferred to surroundings; heat loss centre detects rise in temperature and stimulates responses in skin; such as vasodilation and sweating; so that more heat is transferred to surroundings; **[6]**

23 Nervous system

23.1 **(a)** A = nerve ending/dendrite; which are sensitive to a stimulus/detect change/act as receptor; **[2]**
B = nerve endings/dendrites; connect with a motor neurone; **[2]**
C = cell body; containing cell body and organelles; **[2]**
D = nucleus; contains genetic material; **[2]**
E = axon; transmits impulse away from cell body; **[2]**

(b) on motor neurone arrow is downwards from C to E; arrow on sensory neurone is downwards from A to B; **[2]**

(c) neurone = nerve cell; stimulus = a condition or change in a condition which is detected by a receptor; receptor = specialised sensory neurone which can detect specific conditions or changes in conditions; neuroglial cell = non-neurone cells within nerve tissue, which provide nutrients, may produce the covering of nerves, etc; Schwann cell forms the myelin sheath around a neurone; **[5]**

(d) primary receptor is formed from one sensory cell and is sensitive to a particular stimulus; a secondary receptor is made of more than one cell; **[2]**

23.2 nervous system forms a link between external surroundings and cells; and links intercellularly; formed of neurones; may be grouped as nerves; sensory neurones linked to receptors which detect specific stimuli; connecting neurones in brain and spinal cord; brain and spinal cord make up central nervous system; nervous tissue arising from spinal nerves makes up peripheral nervous system; motor neurones contact effector organs; such as muscles and glands; stimulating effector organs brings about change; autonomic nervous system and automatic responses; reflexes; **[10]**

23.3 **(a)** by active transport/ion pumps; **[1]**

(b) differential permeability/membrane more permeable to K^+/less permeable to Na^+ **[1]**

(c) outer surface of membrane positive in comparison to inside of membrane; **[1]**

(d) reverse it; by opening to allow Na^+ to move in; **[2]**

23.4 A = depolarisation; B = action potential; C = repolarisation; D = refractory period; **[4]**

23.5 **(a)** no impulse transmitted until a big enough stimulus is supplied; **[1]**

(b) once threshold is reached, a bigger stimulus does not cause a bigger response; **[1]**

(c) response of a whole nerve is different because once threshold is reached, a bigger stimulus can trigger impulses in a larger number of neurones; so a greater response overall is seen; **[2]**

23.6 **(a)** interface between two nerve cells; **[1]**

(b) chemical which moves across synapse/ transfers impulse to next neurone; **[1]** e.g. adrenaline/noradrenaline/acetylcholine; **[1]**

(c) excitatory neurotransmitter reinforces/ triggers impulse in next neurone; inhibitory neurotransmitter blocks impulse in next neurone; **[2]**

(d) receptors on sensory neurones respond to stimulation decreasingly; when continually stimulated; so that body does not over-react to stimulus such as touch of clothing against the skin; **[3]**

(e) agonistic drugs reinforce action of natural neurotransmitter; e.g. tranquiliser such as Prozac; antagonistic drugs reduce action of neurotransmitter; eg beta-blocker; **[4]**

23.7 **(a)** Sketch shows these structures labelled: presynaptic structure (shape not too important) with presynaptic membrane; postsynaptic membrane at muscle interface, with underlying muscle layer shown; several mitochondria; a myelinated axon; vesicles with neurotransmitter; **[5]**

(b) in both cases neurotransmitters; from presynaptic component cause response; synaptic cleft exists; mitochondria are organelles which transfer energy; generator potential produced; in postsynaptic membrane; postsynaptic component is different: either a motor neurone or muscular fibres; **[4]**

(c) motor neurone disease; **[1]**

(d) some muscle fibres within whole muscle are contracted at any one time; exercise/muscle use; **[2]**

23.8 **(a)** increasing gradually up to 1 millisecond; increasing sharply from 1–1.8 ms; peak at +40 mv; decreasing sharply from 1.8–2.6 ms where it drops below original level slightly; increasing to –60 mV; and remaining constant; **[any 5 points in correct order]**

(b) sodium gates open; and Na^+ ions flood in across membrane; increasing positive ions inside membrane in relation to outside; **[2]**

(c) delay = 0.6 ms; is time taken for neurotransmitters; to diffuse across gap; **[3]**

(d) increases speed; acts as a stimulant; **[2]**

23.9 **(a)** **A** cerebral hemisphere; **B** cerebellum; **C** hypothalamus; **D** pituitary gland; **E** medulla/medulla oblongata; **[5]**

(b) studying brains of other animals; experiments with human volunteers; from cases of abnormal development; from injury cases; **[3]**

(c) See diagram at bottom of page;

(d) show sensory component including receptors **[2]** show brain and spinal cord (CNS) as processing unit; show motor component as muscles and glands; name peripheral nervous system as part which carries impulses from CNS to effectors; **[3]**

23.9 **(c)**

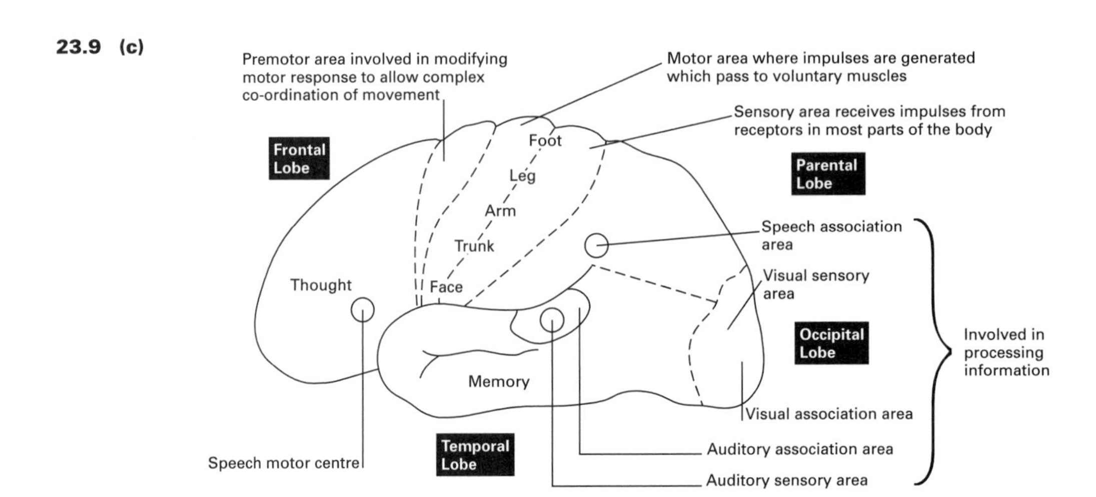

[3]

23.10

Statement	True	False
Sensory neurones enter the spinal cord through the dorsal root	✓	
A dorsal root ganglion is a collection of motor neurone cell bodies		✓
The white matter is mostly made up of collagen tissue		✓
The grey matter is composed of relay neurones and synapses	✓	
Sensory nerves link the spinal cord to the brain	✓	
An unconditioned reflex is modified by experience		✓
Swallowing is an unconditioned reflex	✓	

[7]

23.11 roughly spherical sense organ; three layered wall; inner layer contains light sensitive receptors; which send visual information in form of impulses to brain; at front of eye is ring of muscle called ciliary muscle; contraction of ciliary muscle; causes ligaments which hold lens to slacken; this allows adjustment in thickness of lens; and hence focal length/point; cornea is outer layer which helps in converging light rays; **[10]**

23.12 **(a)** bipolar neurone = A; rod = C; cone = B; optic nerve = D; **[4]**

(b) cones operate singly; so need a large enough stimulus for impulses to be triggered; rods converge in small groups; so their effect adds up/summates; and they don't need as much light for impulses to be triggered; **[5]**

(c) inability to distinguish certain colours e.g. red/green; genetic cause; **[2]**

23.13 sight/smell of food detected by receptors; of sensory neurones and results in saliva glands; making more saliva; nerve cells in stomach detect presence of food; and message is processed by brain; resulting in glands; in duodenum producing hormones; secretin and cholecystokinin; so both systems are involved; cholecystokinin causes muscle of bile duct; to contract so bile enters duodenum; and secretin stimulates liver to produce bile; **[10]**

23.14 **(a)** 10 mg/100 ml **[2]** (1 for unit)

(b) brain activity increases in areas associated with lowering inhibitions and boosting confidence; **[2]**

(c) motor coordination affected/sedated; responses slowed; and confidence rises; **[2]**

(d) around 80 mg/100 ml; **[1]**

(e) low level alcohol stimulates NMDA receptors; which increases brain activity; higher levels of alcohol stimulate GABA receptors; which sedate nerve responses; **[4]**

(f) glucose needed for cellular respiration; otherwise not as much energy can be transferred to muscle fibres to allow contraction; **[2]**

24 Effectors: muscles and glands

24.1 **(a)** moving limbs by pulling on bones; peristaltic movements in the gut/along ducts; contraction in circular muscles to open and close apertures e.g. iris, sphincter muscles; **[3]**

(b) control/reduce/alter glucose level in blood; cause lactation; cause muscle contraction as in labour; cause growth changes e.g. at puberty or in childhood; control metabolic rate; control reabsorption of water in kidneys; or other; **[3]**

(c) liver composed of more than one type of tissue – depends on the tissue stimulated; **[1]**

24.2 1 cardiac muscle; 2 motor neurones of peripheral nervous system; 3 intrinsic: sinoatrial node initiates heart beat; 4 spindle shaped fibres, unbanded; 5 fast to act and fast to fatigue; voluntary; 6 walls of organs and ducts e.g. gut, uterus, artery walls, iris; 7 large muscles in limbs, attached to bones; 8 the heart; **[8]**

24.3

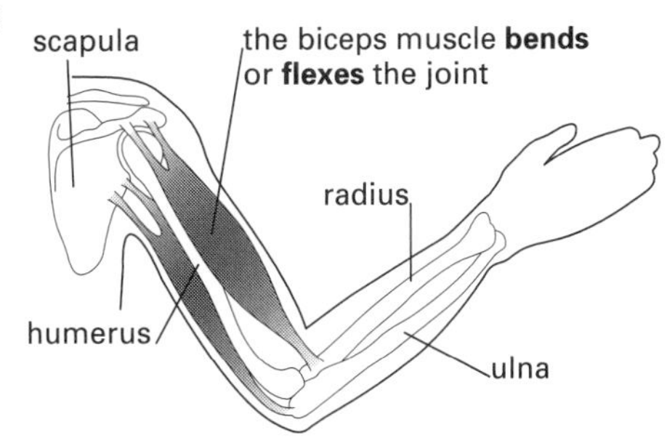

Diagram needs to show reasonable likeness: humerus, radius and ulna, biceps and triceps, muscle attachments by ligaments and labelled **[5]**
account to include: muscle attached to bone by ligaments which do not stretch; muscle contracts, shortening in length and pulling on bone, biceps flexes arm and triceps straightens it/antagonistic pair of muscles **[5]**

24.4 **(a)** sketch shows two bones, one either side of a simple joint; with muscle attached either side of the joint; a joint (such as elbow or knee) acts as a fulcrum; the muscle pulls on a bone/acts as the effort; the mass of the limb is the load; **[5]**
(b) shorter limbs are easier to control; less mass to move; **[2]**
(c) less muscle mass; not as strong/powerful **[2]**

24.5 **(a)** bone: strong , hard, can resist compression; provides support; **[2]**
(b) cartilage: hard, flexible/elastic good shock absorber; hard-wearing; **[2]**
(c) ligament: strong and tough but flexible; hold joints together; **[2]**
(d) tendon: strong but inelastic; connect muscles to bones; **[2]**

24.6 contraction; actin; calcium; bridges/head; ATP; phosphocreatine; **[6]**

24.7 **(a)** slow twitch: contract and fatigue slowly **[2]**
fast twitch: contract and fatigue rapidly **[2]**
(b) oxygen carrier in muscle tissue; **[1]**
(c) slow twitch; **[1]**
(d) oxygen needed to transfer ATP during respiration; which is energy source for contraction of muscle; **[2]**
(e) increases proportion of fast twitch fibres; **[1]**
(f) stronger/denser bones; damaged joints/stretched ligaments; more supple; **[1]**
(g) cardiovascular; **[1]**
(h) eat high starch/carbohydrate diet; storage of glycogen in muscles for energy release during long runs; **[2]**

24.8 **(a)** stretch receptor or proprioceptor; **[1]**
(b) via sensory nerve of central nervous system; **[1]**
(c) force of contraction of muscle fibre increases; so muscle can do more work without contracting in length; **[2]**
(d) proportion of muscle fibres which are contracted in a generally relaxed muscle; **[2]**
(e) muscle is not completely relaxed; **[1]**
(f) holds body upright/specified position; **[1]**

24.9 **(a)** 0.6 l min^{-1}; **[1]**
(b) rises steeply between 1200 and 1201 h; rises less steeply between 1201 and 1202 h; slows during third minute to almost constant consumption; **[3]**
(c) increase in ventilation rate; and tidal volume/volume of breath; **[2]**
(d) increases; **[1]**
(e) curve drops sharply over the next two minutes; and then levels off to about same level as at start; (**see diagram below**) **[2]**
(f) lactate builds up in muscle; when oxygen deficiency occurs; need to get enough oxygen to respire remaining lactate after exercise finishes; **[3]**

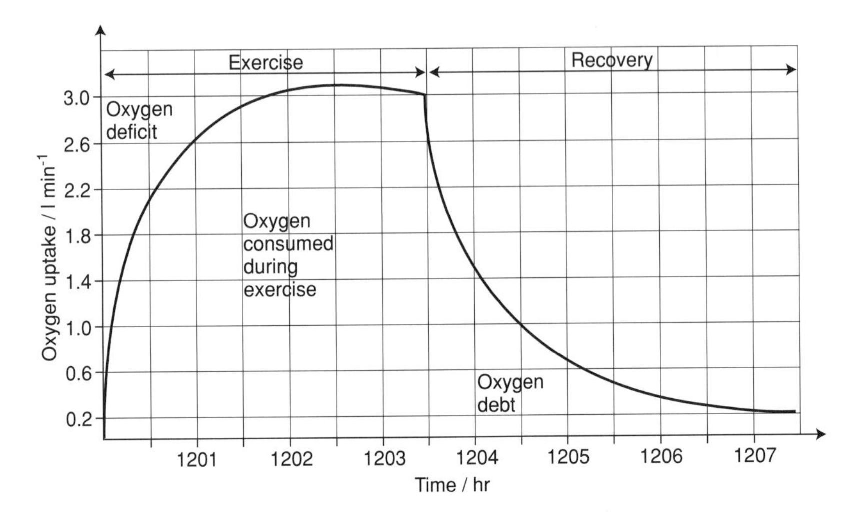

[3]

24.10 **(a)** produced in response to stress/inflammation/allergy to counteract effects; **[1]**

(b) quantitative: X-ray/level of cortisol in urine **[2]**

qualitative: talking to girls about how concerned they were about body mass; **[1]**

(c) quantitative; because it can be verified by repeating the work; is measurable; more reliable; does not rely on opinion; **[2]**

(d) genetic factor; level of exercise; dietary factor such as lack of calcium or phosphate; **[3]**

(e) ethnic group; exercise level; diet; family medical history; **[2]**

(f) what is true for college girls is not necessarily true for pre-teen age group as there are many other variables, including effects of puberty **[2]**

(g) weak bones fracture more easily; especially in elderly; breaks can lead to other complications if recovery is slow; bones as act mineral store; **[3]**

(h) exercise; mineral supplements/alter diet; **[2]**

24.11 (Use standard text to help with this)

24.12 hormone e.g. adrenaline; interacts with receptor site; in surface cell membrane; second messenger e.g. AMP; activates enzymes in cell; may be many stages/cascade effect; hormone e.g. oestrogen; and receptor may move into cell; and cause transcription; may bind to receptor site on membrane and alter permeability; of membrane **[10]**

24.13 **(a)** very similar; based on hydrocarbon structure; only vary in one alkyl (CH_3) group; **[3]**

(b) steroids; **[1]**

(c) membrane surrounding a vesicle has partially lipid structure; so cannot prevent movement of hormone which is lipid soluble; **[2]**

(d) these are active compounds; chemical composition can be altered slightly; to give a compound which also fits receptor sites; but has different effect; **[2]**

24.14 **(a)** reduces volume; **[1]**

(b) ADH (antidiuretic hormone); **[1]**

(c) increases permeabillity of collecting ducts; allowing more water back out into blood; **[2]**

(d) insulin intravenously; controlled diet/calorie intake; **[2]**

(e) occurs during pregnancy; **[1]**